Our Energy Future

Rapid changes in energy production and consumption are having major socioeconomic implications for the communities of rural America. Technological developments in horizontal drilling and hydraulic fracturing (fracking) have significantly increased domestic energy production. These changes have implications for nuclear and renewable energy as well as efforts to improve energy conservation and efficiency.

In this time of rapid change in the energy industry, policy decisions on energy issues will have profound implications for all Americans. This is especially true for rural communities where impacts are experienced most directly. Thus, the time is appropriate for a careful exploration of the socioeconomic implications of our energy future.

The purpose of this text is to present timely and scientifically sound information on energy policy and the socioeconomic aspects of energy production and consumption with a focus on rural areas. The book presents the latest research by top scholars with the goal of clarifying options and providing the basis for informed policy decisions.

Don E. Albrecht, PhD, is Director of the Western Rural Development Center and Associate Director of the National Agricultural and Rural Development Policy Center (NARDeP). Dr Albrecht has extensive research experience and numerous publications on issues confronting rural America. He has worked closely with policy makers and other leaders on approaches to address these issues.

Routledge Studies in Energy Policy

Our Energy Future
Socioeconomic Implications and Policy Options for Rural America
Edited by Don E. Albrecht

Our Energy Future

Socioeconomic Implications and Policy Options for Rural America

Edited by
Don E. Albrecht

Routledge
Taylor & Francis Group
LONDON AND NEW YORK

First published 2015
by Routledge
2 Park Square, Milton Park, Abingdon, Oxfordshire OX14 4RN

and by Routledge
711 Third Avenue, New York, NY 10017

First issued in paperback 2016

Routledge is an imprint of the Taylor & Francis Group, an informa business

British Library Cataloguing in Publication Data
A catalogue record for this book is available from the British Library

Library of Congress Cataloging in Publication Data
Our energy future : socioeconomic implications and policy options for rural America / [edited by] Don E. Albrecht.
pages cm -- (Routledge studies in energy policy)
Includes bibliographical references and index.
1. Energy policy--United States. 2. Energy development--United States. 3. Land use, Rural--United States. 4. Rural development--United States. I. Albrecht, Don E.
HD9502.U52O82 2014
333.790973--dc23
2014004358

ISBN 13: 978-1-138-24075-9 (pbk)

ISBN 13: 978-1-138-78411-6 (hbk)

Typeset in Sabon
by Saxon Graphics Ltd, Derby

Contents

Figures

Tables

Contributors

Don E. Albrecht, PhD, is Director of the Western Rural Development Center and Associate Director of the National Agricultural and Rural Development Policy Center (NARDeP). Dr Albrecht has extensive research experience and numerous publications on issues confronting rural America. He has worked closely with policy makers and other leaders on approaches to address these issues.

Amanda Weinstein is an Assistant Professor at the University of Akron, USA, in the Department of Economics. She is a member of the Council of Regional Economic Policy Advisors for the Fund for Our Economic Future and a Faculty Associate of the Regional Innovation Institute at the University of Akron.

Mark Partridge is the Swank Chair of Rural-Urban Policy at the Ohio State University, USA, and a Professor in the AEDE Department. He has affiliations with the University of Alberta, Canada, the University of Toronto, Canada, and the University of Saskatchewan, Canada. He has published over one hundred scholarly papers about his work.

Susan Christopherson is a Professor and Chair of the Department of City and Regional Planning at Cornell University, USA. An economic geographer committed to the integration of scholarly work and public engagement, her diverse research interests center on industry location decisions and regional, state, national and international economic development policy.

Ned Rightor is President of New Economy Dynamics LLC, a research and consulting firm focused on workforce development and economic development projects throughout the northeastern USA. Since 2010, in collaboration with Cornell University researchers, he has been conducting multi-state studies of the social and economic consequences of unconventional oil and gas development.

Bruce A. McCarl is Regents Professor and Distinguished Professor of Agricultural Economics at Texas A&M University, USA. His recent research efforts involve policy analysis (mainly in climate change, climate change adaptation, climate change mitigation, water economics and biosecurity) as well as the application of quantitative methods to such analyses. He teaches applied mathematical programming and risk analysis. He was part of the 2007 Nobel Peace Prize winning Intergovernmental Panel on Climate Change.

Marta Wlodarz is a post-doctoral researcher in the Centre for Energy Policy and Economics at the ETH Zürich, Switzerland. She has research experience in the area of natural resources economics, biofuel economics and climate change economics. She wrote several climate and energy policy briefs, book chapters and articles. At the moment, she researches the macroeconomic implications of various energy and carbon policies for the economic growth and environment of European countries.

Beth-Anne Schuelke-Leech, PhD, is an Assistant Professor in the John Glenn School of Public Affairs at Ohio State University, USA. Her research focuses on the connections between engineering, business, innovation and public policy. In addition to her scholarship, she worked as a professional engineer for over a decade.

Harry de Gorter is a Professor in the Dyson School of Applied Economics and Management, Cornell University, USA. His research focus is on agricultural trade policy with recent work on biofuels and agricultural trade reform and the Doha Development Agenda, especially the impact of subsidies and protection on developing countries.

Dusan Drabik is an Assistant Professor of Agricultural Economics at Wageningen University in the Netherlands, and a Research Fellow at LICOS, KU Leuven, Belgium. Dusan's research focus is on the analysis of agricultural and environmental policies, especially on biofuel policies in the United States, the European Union and developing countries.

Richard S. Krannich is Professor of Sociology at Utah State University, USA. He has served as President of the Rural Sociological Society and Executive Director of the International Association for Society and Natural Resources. His research examines social implications of environmental change and resource use in rural communities of the Intermountain West.

Peter G. Robertson is a doctoral student in the Department of Environment and Society at Utah State University, USA. His research investigates the social effects of renewable energy development in the American West. He is also interested in understanding the political, economic and social barriers to the future implementation of renewable energy systems.

Shawn K. Olson is a doctoral student in Sociology at Utah State University, USA. Her research investigates community perceptions of and social conflicts over the siting of industrial-scale renewable energy facilities in the rural American West. She is also interested in the political economic dimensions of renewable energy development nationwide.

Matteo Muratori is a Research Associate Engineer at the Ohio State University, USA, where he conducts research on modeling integrated energy systems, sustainable mobility, dynamic energy management techniques and the evaluation of energy policies, including technological and economic implications of the integration of smart technologies.

Acknowledgements

This book is the result of a project initiated by National Agricultural and Rural Development Policy Center (NARDeP). NARDeP is organized by the Regional Rural Development Centers and funded by the USDA National Institute of Food and Agriculture (NIFA) under a competitive grant. The goal of NARDeP is to develop and deliver timely policy-relevant information about complex and contentious rural development policy issues. A NARDeP Policy Brief has been developed for each of the chapters in this book.

1 U.S. Energy Production and Consumption

An Overview

Don E. Albrecht

Historically, the primary sources of energy for agriculture, transportation, industry and other human endeavors were efforts of humans and their domesticated animals. Wood and other forms of biomass were the major sources of heat for cooking and home heating. Limitations imposed by these energy forms suppressed productivity and living standards. Transportation was slow, work was tedious and productivity minimal.

Much changed with the start of the Industrial Revolution. In the main, the Industrial Revolution was simply a result of the growing capacity of humans to harness and utilize fossil fuels as an energy source. Fossil fuels were formed from organisms that lived long ago and over time were transformed into coal, oil, natural gas or other products that burn easily and release energy as they do so. The increased capacity of humans to use fossil fuels as an energy source has completely transformed the way of life for millions of people and has resulted in rapid economic growth. Fossil fuels power cars, trucks, trains and planes, and they have dramatically increased transportation speed and capacity. Fossil fuels are used to operate farm equipment and factories, and they have greatly increased the work capacity of individuals. Fossil fuels heat and cool our homes and provide electricity to keep the lights on and to operate our televisions, radios and computers with the flip of a switch. Additionally, fossil fuels are used extensively in plastics, fertilizer, pesticides and many other products of our modern world.

Table 1.1 provides data on international energy consumption. This table shows that while the U.S. had only 4.6 percent of the world's population in 2006, Americans consumed 21.1 percent of the energy. Per capita energy consumption in the United States is far greater than in other places around the world. The average American consumes twenty-one times more energy than the average person from either Africa or India, six times more than the average Chinese, and twice as much as the average European where incomes in many countries are as high or higher than in the United States. Energy consumption is strongly related to Gross Domestic Product. Generally, countries that consume the most energy have the highest GDPs. However, Table 1.1 shows that Europeans are more efficient than the U.S. in that they produce more with less energy.

Table 1.1 International Energy Consumption in 2006

Region	*Total Population (Millions)*	*Percent*	*Energy Consumption (Quadrillion BTUs)*	*Percent*	*Per Capita Energy Consumption (Million BTUs)*	*BTUs Per 2000 U.S. Dollars Using Purchasing Power Parities*
North America	438.60	6.7	121.182	25.7	276.2	8,933
United States	298.36	4.6	99.856	21.1	334.6	8,841
Central and South America	456.21	7.0	24.176	5.1	53.3	7,283
Europe	593.64	9.1	86.422	18.3	146.2	6,536
Eurasia	284.60	4.4	45.880	9.7	160.8	20,137
Middle East	190.46	2.9	23.806	5.0	127.2	13,709
Africa	924.51	14.1	14.495	3.1	15.9	6,651
Asia & Oceania	3,649.36	55.8	156.313	33.1	42.8	9,241
World Total	6,537.37	100.0	472.274	100.0	72.4	8,874

Source: Energy Information Administration, U.S. Department of Energy.

Table 1.2 Energy Production and Consumption by Source in the United States, 1950–2011 (Quadrillion BTU)

	Fossil Fuels		*Nuclear*		*Renewable*		*Total*	
Year	*Production*	*Consumption*	*Production*	*Consumption*	*Production*	*Consumption*	*Production*	*Consumption*
1950	32.563	31.632	0.000	0.000	2.978	2.978	35.540	34.616
1955	37.364	37.410	0.000	0.000	2.784	2.784	40.148	40.208
1960	39.869	42.137	0.006	0.006	2.929	2.929	42.804	45.087
1965	47.235	50.577	0.043	0.043	3.398	3.398	50.676	54.017
1970	59.186	63.522	0.239	0.239	4.070	4.070	63.495	67.838
1975	54.733	65.357	1.900	1.900	4.687	4.687	61.320	71.965
1980	59.008	69.828	2.739	2.739	5.428	5.428	67.175	78.067
1985	57.539	66.093	4.076	4.076	6.084	6.084	67.698	76.392
1990	58.560	72.332	6.104	6.104	6.041	6.041	70.705	84.485
1995	57.540	77.259	7.075	7.075	6.558	6.560	71.174	91.029
2000	57.366	84.731	7.862	7.862	6.104	6.106	71.332	98.814
2005	55.056	85.794	8.161	8.161	6.229	6.242	69.428	100.282
2011	60.601	79.779	8.259	8.259	9.236	9.135	78.096	97.301

Source: Energy Information Administration, U.S. Department of Energy.

Table 1.2 provides data on energy production and consumption in the United States from 1950 to 2011. Most significantly, this table shows that energy consumption increased greatly during this time period. In 2011 Americans were consuming about three times more energy than in 1950. Of course, some of this increase is a consequence of population growth. The extreme dependence of the United States on fossil fuels is also apparent. In 2011, 82.1 percent of energy consumed in the United States was from fossil fuels, with only 8.5 percent from nuclear power and 9.4 percent from renewable energy sources.

While the benefits of fossil fuel development are extensive, nevertheless there are major concerns associated with our continued heavy reliance on fossil fuels. Included among the issues associated with fossil fuel dependence are: (1) the inevitable depletion of finite fossil fuel resources; (2) problems resulting from reliance on foreign energy supplies; (3) environmental problems associated with fossil fuel extraction and transportation; (4) emission of climate-changing greenhouse gasses; and (5) socioeconomic and community implications. Because of these concerns, our dependence on fossil fuel must be reduced (Geri and McNabb 2011; Lester and Hart 2012; Sioshansi 2011). This chapter begins with an overview of fossil fuel production and consumption. A discussion is then provided of the relationship between fossil fuels and the five issues mentioned above. The chapter concludes with a discussion of the options available for a transition away from fossil fuel dependence. For several of the most significant or recently emerging issues, more detail is provided in later chapters of this book.

Fossil Fuel Production and Consumption

Table 1.3 provides a more detailed examination of fossil fuel production and consumption in the United States from 1950 to 2011. Coal (24.6 percent) and natural gas (31.1 percent) provided more than one half of the energy from fossil fuels consumed in the U.S. in 2011. However, almost half (44.3 percent) of the fossil fuel energy consumed in the U.S. in 2011 was from petroleum. The U.S. has extensive coal and natural gas reserves, and it is evident from Table 1.3 that current production of these resources is sufficient to meet present demand. On the other hand, petroleum production falls far short of consumption, requiring substantial imports.

Petroleum

Petroleum is the single most important energy source in the world today. Transportation is the most prominent user of petroleum products, most of which are derived from crude oil. In the U.S., the transportation sector uses about 71 percent of the petroleum consumed to fuel approximately 250 million vehicles. To put U.S. petroleum consumption into perspective,

Table 1.3 Fossil Fuel Production and Consumption by Type in the United States, 1950–2011 (Quadrillion BTU)

	Coal		*Natural Gas*		*Petroleum*	
Year	*Production*	*Consumption*	*Production*	*Consumption*	*Production*	*Consumption*
1950	14.060	12.347	6.233	5.968	12.270	13.315
1955	12.370	11.167	9.345	8.998	15.650	17.255
1960	10.817	9.838	12.656	12.385	16.396	19.919
1965	13.055	11.581	15.775	15.769	18.404	23.246
1970	14.607	12.265	21.666	21.795	22.913	29.521
1975	14.989	12.663	19.640	19.948	20.103	32.731
1980	18.598	15.423	19.908	20.235	20.503	34.205
1985	19.325	17.478	16.980	17.703	21.233	30.925
1990	22.488	19.173	18.326	19.603	17.746	33.552
1995	22.130	20.089	19.082	22.671	16.329	34.438
2000	22.735	22.580	19.662	23.824	14.969	38.262
2005	23.185	22.797	18.556	22.565	13.297	40.388
2011	22.181	19.643	23.506	24.843	14.914	35.283

Source: Energy Information Administration, U.S. Department of Energy.

according to the Energy Information Administration, about 378 million gallons of gasoline are consumed each day – and this is not counting diesel, jet fuel, home heating oil, or petroleum used in the production of asphalt, plastics and other products.

Among the major problems associated with crude oil in the U.S. are inadequate supplies and the consequent depletion of existing resources. As of 2011, less than 10 percent of global crude oil production comes from the United States. U.S. crude oil production reached a peak in 1970 and then declined rapidly until a recent resurgence (Table 1.3). Until about 2008, declining production was rather pronounced. By 2005, after years of declining production and growing consumption, only 32.9 percent of the petroleum consumed in the United States was produced there.

Table 1.4 makes the reasons for declining U.S. oil production even more apparent. This table shows that the overall U.S. oil production increased until 1970. This increase was associated with the discovery of new oil fields and improved technology that allowed average well production to increase steadily until it reached a high of 18.1 barrels per well per day. Since that time, per well production has declined significantly as oil resources have become depleted and wells in less productive locations have been brought on line. By 2008, per well production was averaging only 9.4 barrels per day. Again, there have been increases in recent years.

The severity of the decline in U.S. oil production was initially offset by increased production from Alaska. In the 1960s, what proved to be the largest oil field in North America was discovered at Prudhoe Bay on the North Slope of Alaska. This oil field originally contained an estimated twenty-five billion barrels of oil. In 1975 construction began on an

Table 1.4 Crude Oil Production by Location and Average Productivity per Well in the United States, 1955–2011

	Location			
	48 States	*Alaska*	*Total*	*Average Well Productivity*
Year	*(Thousand Barrels Per Day)*			*(Barrels Per Day)*
1955	6,807	0	6,807	13.0
1960	7,034	2	7,036	11.9
1965	7,774	30	7,804	13.2
1970	9,408	229	9,637	18.1
1975	8,183	191	8,374	16.8
1980	6,980	1,617	8,597	15.7
1985	7,146	1,828	8,974	13.9
1990	5,582	1,773	7,355	12.2
1995	5,076	1,484	6,560	11.4
2000	4,851	970	5,820	10.9
2005	4,314	864	5,178	10.4
2011	5,090	572	5,662	10.6

Source: Energy Information Administration, U.S. Department of Energy.

eight-hundred-mile pipeline to transport oil from Prudhoe Bay to Valdez, Alaska, the northernmost ice-free port in North America. The pipeline cost an estimated $8 billion to build and opened in 1977. Since that time over fifteen billion barrels of oil have been successfully transported. With the pipeline in place Alaskan oil production reached a peak in 1985, but it has since declined sharply with the depletion of the Prudhoe Bay oil fields.

In recent years U.S. oil production has been enhanced by the development of new technologies. Horizontal drilling and hydraulic fracturing (fracking) have made it economically feasible to produce "light tight" oil from shale formations. This same technology has also dramatically impacted natural gas production (which will be discussed later). Utilization of these "unconventional" strategies has allowed development to occur in the Bakken and other shale formations. The Bakken Field is centered in North Dakota, extending into Montana as well as Saskatchewan and Winnipeg in Canada. As a consequence of the advent of fracking, North Dakota oil production increased from 29 million barrels in 2003 to 153 million barrels in 2012. As a consequence of recent production increases and consumption declines, by 2011 the U.S. was able to produce 42.3 percent of the petroleum it consumed. The Energy Information Administration projects that fracking will greatly increase U.S. energy production and thus reduce the need for energy imports. In fact, the International Energy Agency has forecast that the U.S. will become the world's biggest oil producer by 2020, and a decade later it will become energy independent.

Coal

Known coal deposits in the U.S. contain more energy than all the world's known oil reserves combined. Not only is coal relatively plentiful, it is also cheap to produce. Coal accounted for nearly one half of the increase in global energy use over the past decade, largely because of increased coal development in China (International Energy Agency 2011). Currently coal is used to generate about half of the electricity consumed in the U.S. While U.S. coal supplies are relatively plentiful, two factors will inhibit coal use in the future. First, coal produces more pollution and climate-changing greenhouse gasses per unit of energy generated than other energy alternatives (e.g. Jaramillo et al. 2009). For example, coal produces more than twice as much carbon dioxide per unit of electricity generated than natural gas. Second, given current technology, coal cannot be used efficiently in the transportation sector. As a consequence of these disadvantages, coal production and consumption has declined slightly in recent years. Given climate change concerns, it is very likely that the key innovation for coal use in the future will be the development of carbon capture and storage or other "clean coal" technologies that will allow coal to be used in a cost-effective manner while also protecting the environment.

Table 1.5 presents data showing coal production in the United States from 1950 to 2011. This table indicates that coal production has increased substantially since 1960. Much of this increased coal production came following the OPEC embargo of the 1970s. This embargo revealed the extent of dependence of the U.S. on foreign energy sources, and thus made our vulnerability evident. The result was significant efforts to increase domestic production to help achieve energy independence.

Table 1.5 also shows that down the years there have been major changes in mining methods and the location of coal production in this country. In 1950, 75 percent of coal produced in the United States was from underground mines, and most (93.6 percent) was taken from mines east of the Mississippi. The most significant coal mining areas in the middle of the last century were in the Appalachian regions of West Virginia, eastern Kentucky and western Pennsylvania. In the decades since, substantial amounts of coal have continued to be mined using underground methods in the eastern states. However, there have been dramatic increases in the amount of coal obtained from surface mines and from mines in the west. With modern technology surface mining is far cheaper, and western coal deposits are more amenable to surface mining. With these techniques, massive machines can be used to remove the soil and other materials covering the coal, extract the coal and then transport it to power generating facilities. By 2011, 68.4 percent of U.S. coal was obtained from surface mines and 58.3 percent was from the west. Wyoming is now the nation's leading coal-producing state, and Campbell County, Wyoming is the nation's top coal-producing county with more than one million tons of coal per day being mined there. In Campbell County and the Powder River Basin, coal seams from 65 to 100 feet thick are located near the surface and are thus relatively easy and cheap to mine. In some cases power plants are located at the mouth of the mine, and coal is conveyed from the mine to the plant where electricity is generated. Transmission lines then carry this power to distant cities. At other mines the coal is loaded onto trains that carry the coal to power plants located throughout the country.

Natural Gas

Natural gas consists of methane and other hydrocarbons that are naturally occurring and are found in deep underground rock formations. At one time, natural gas was considered a relatively worthless by-product of oil production. If markets could not be found near oil wells, natural gas was simply burned off at the oil field. In recent years, however, markets for natural gas have expanded greatly and gas-to-liquid technologies make transporting natural gas economically feasible.

Today, natural gas is a tremendously valuable resource that is widely used for heating, electricity generation, and increasingly as a transportation fuel. Table 1.6 provides data on natural gas production in the United States

Table 1.5 Coal Production in the United States by Mining Method and Location, 1950–2011 (Million Short Tons)

	Mining Method		*Location*		
Year	*Underground*	*Surface*	*East of Mississippi*	*West of Mississippi*	*Total*
1950	421.0	139.4	524.4	36.0	560.4
1955	358.0	132.8	464.2	26.6	490.8
1960	292.6	141.7	413.0	21.3	434.3
1965	338.0	189.0	499.5	27.5	527.0
1970	340.5	272.2	567.8	44.9	612.7
1975	293.5	361.1	543.7	110.9	654.6
1980	337.5	492.2	578.7	251.0	829.7
1985	350.8	532.8	558.7	324.9	883.6
1990	424.5	604.6	630.2	398.9	1,029.1
1995	396.6	636.9	544.8	488.7	1,033.5
2000	373.6	700.0	507.5	566.1	1,073.6
2005	368.6	762.9	493.8	637.7	1,131.5
2011	345.5	748.8	455.8	638.5	1,094.3

Source: Energy Information Administration, U.S. Department of Energy.

Table 1.6 Natural Gas Production in the United States by Type of Well 1950–2011 (Million Cubic Feet)

Year	*Gas Wells*	*Oil Wells*	*Coal Bed Wells*	*Shale Gas Wells*	*Total*
1950	NA	NA	0	0	8,479,650
1960	NA	NA	0	0	15,087,911
1970	18,594,658	5,191,795	0	0	23,786,453
1980	17,572,526	4,297,166	0	0	21,869,692
1990	16,053,566	5,469,055	0	0	21,522,622
2000	17,726,056	6,447,820	0	0	24,173,875
2007	14,991,891	5,681,871	1,999,748	1,990,145	24,663,656
2011	12,291,070	5,907,919	1,779,055	8,500,983	28,479,026

Source: Energy Information Administration, U.S. Department of Energy.

from 1950 to 2011. Natural gas production increased rapidly from 1950 to 1970, and then leveled off for the next thirty years. In the past decade natural gas production has increased significantly. This is largely a consequence of recent developments in production technology, which include horizontal drilling and hydraulic fracturing (fracking). These technologies have greatly increased the amount of natural gas that can be economically produced by allowing access to gas that is trapped in shale formations. These unconventional natural gas technologies have been a real "game changer." Table 1.6 shows that shale gas production has gone from zero in 2000 to about 30 percent of all production in 2011. Currently, the highest producing shale formations include Haynesville (Louisiana and Texas), Marcellus (Pennsylvania, West Virginia, New York and Ohio),

Barnett (Texas), and Fayetteville (Arkansas). Greatly increased supplies made available by fracking have also resulted in substantial reductions in natural gas prices, with widespread implications. Chapter Two in this book examines the economic implications of these new technologies, while Chapter Three explores social and community implications.

Issues with Fossil Fuel Dependence

Depletion of Finite Fossil Fuel Resources

Fossil fuels are not a renewable resource. They developed over geological time, and when they are used, they are gone. The fossil fuels we use today will simply not be available for our children or grandchildren. Resource depletion does not occur in a manner whereby the resource exists one day and is gone the next. Rather, a process of "economic decline" occurs. As demand grows, extensive efforts are made to locate additional supplies and extract them efficiently. Eventually, as nonrenewable resources are used, supplies necessarily dwindle. The cheapest supplies are diminished first and then other more expensive and difficult to obtain supplies are brought into use. As this occurs, costs inevitably increase. Higher costs then put a damper on demand. Over time, increasingly expensive supplies will result in continually higher prices and therefore in diminishing demand. Eventually, costs will become so high that incentives to find alternatives will increase and the use of the base resource will virtually cease. New discoveries and technological breakthroughs can slow this process, but the end result is inevitable.

Ever-increasing resource use obviously speeds up resource depletion processes. The world's population is growing and people are getting wealthier, and both of these factors lead to increased consumption. Like Americans, as people around the world become wealthier, they use their higher incomes to improve their standard of living. These changes involve the consumption of vastly greater amounts of energy (McKibben 2007). For example, as incomes in China have increased, the number of cars purchased and the number of miles driven annually in China have also increased rapidly. The number of cars purchased annually in China has already overtaken the same statistic for the United States (Yergin 2011).

The easiest approach for dealing with resource depletion in the U.S. is to import foreign energy. However, this strategy generates problems of its own, which will be discussed later. Furthermore, this is only a short-term solution because even international supplies of a finite resource will eventually become depleted, especially given increased consumption from other parts of the world. Other approaches to address resource depletion problems in the U.S. include increasing the production of substitutes. In the U.S. this has traditionally meant replacing oil with coal and natural gas, both of which are resources with greater remaining domestic supplies.

However, coal and natural gas are also fossil fuels and they face the same problems of resource depletion and greenhouse gas emission as petroleum.

Oil shale is another fossil fuel with massive reserves that may be a critical resource in the future. Oil shale generally refers to any sedimentary rock that contains solid bituminous material called kerogen, which is released as petroleum-like liquids as the rock is heated. This oil shale can be mined and processed to generate oil similar to that which is pumped from conventional wells. While oil shale is found in many places throughout the world, by far the largest known deposits are in the Green River Formation in Colorado, Utah and Wyoming. Estimates of the quantities of recoverable oil from the Green River Formation are three times greater than proven oil reserves in Saudi Arabia. Extracting oil from oil shale is complex because it has to be mined and then heated, and the process is thus more expensive than traditional extraction methods. Furthermore, oil shale production in itself results in greenhouse gas emissions. Additionally, current oil shale recovery technologies utilize extensive amounts of water, which is a problem in the water-scarce west. Because of these problems, current utilization of oil shale is minimal. When oil prices get high enough or if a technological breakthrough similar to the development of fracking occurs, oil shale could play a significant role in meeting the energy needs of the future. However, oil shale is still a fossil fuel and ultimately it will face the same problems as other fossil fuels.

In the final analysis, fossil fuels are a finite resource. While imports, discovering new supplies, using other fossil fuels as substitutes and the development of new technologies may postpone the inevitable, in time, even these resources will become depleted. Whether it happens in fifty, one hundred or one thousand years from now, unless changes are made, our descendants are going to have to deal with the fact that we have left them with a fossil fuel bag that is approaching empty. There is no question that increasing our use of alternative energy sources and improved conservation are essential.

Dependence on Foreign Energy Supplies

Thus far, the U.S. has been able to avoid extreme costs and other problems associated with resource shortages and depletion by importing oil from countries that possess greater supplies. However, foreign oil dependence creates problems of its own. When the U.S. imports foreign oil, this results in massive transfers of wealth from the United States to major oil-producing countries. This leads to significant balance of trade concerns. Furthermore, many oil-exporting countries use oil wealth to maintain non-democratic governments and suppress human rights. Several countries that are major exporters of oil to the U.S. are not considered as shining examples of human rights. Additionally, the U.S. spends vast sums on military power to assure that our interest in the major energy-producing countries is maintained. For these reasons, achieving energy independence has been a primary policy objective of the U.S. for decades because it is recognized that ultimately dependence means vulnerability.

Environmental Problems Associated with Fossil Fuel Extraction and Transportation

By its very nature, the extraction of fossil fuels is environmentally disruptive. Fossil fuels are almost always underground, and soil and rocks must be removed, tunnels dug or wells drilled to reach them. As a consequence, numerous problems have emerged over the years. In April 2010 a massive oil spill occurred in the Gulf of Mexico following an explosion at a BP well. The explosion caused a drill rig to sink at sea, killing eleven workers and injuring seventeen others. An estimated 19,000 barrels of oil per day gushed into the Gulf. By the time the flow could be stopped several months later, more than 120 million gallons of oil had been spewed into the ocean, making this the worst oil spill in world history. In 1989 the sinking of the supertanker *Exxon Valdez* in Alaska's Prince William Sound resulted in 260,000 barrels of oil being spilled. The effects of oil spills for wildlife and biodiversity are extensive and persist for years. An additional concern with fossil fuel mining is that the sought-after resources are mixed with impurities and other unwanted materials that have to be removed. Disposing of these unwanted materials causes a significant problem.

There are also environmental concerns associated with coal mining. For example, the Appalachian coal mining industry has come under fire for what is called "mountain top removal", where the tops of mountains are removed to provide access to coal seams below. With this approach, cheap surface mining techniques can be used. Laws require that land disrupted by mining must be reclaimed. However, reclamation has its limitations, and the disruption and biodiversity losses caused by mining have long-term consequences. Amenity and water quality concerns are also significant. In 2008 the walls of a dam holding 1.1 billion gallons of coal ash crumbled, spilling toxic waste into the town of Kingston, Tennessee and creating the largest industrial spill in U.S. history. Numerous other examples could be cited. Further, environmental concerns such as inducing seismic activity, ground- and waste-water pollution, and the possibility of large-volume methane leaks have recently emerged with respect to increases in the use of fracking technology.

Concerns about the environmental consequences of fossil fuel development have resulted in significant political controversy. For example, in the 1960s another major oil reserve was discovered east of Prudhoe Bay along Alaska's north shore, where an additional estimated ten billion barrels of oil are stored. However, this reserve is located within the boundaries of the Arctic National Wildlife Refuge (ANWR). Questions about the utilization of these oil reserves have resulted in a classic clash of resource development versus preservation. At issue is the fact that the oil is under the primary calving grounds for North America's largest caribou herd, and the region provides a critical habitat for other arctic animals. Consequently, the battle over the ANWR oil has raged for the past thirty years. Other similar battles have been and will be fought.

Climate Changing Greenhouse Gasses

Climate change is perhaps the most serious concern for our continued heavy reliance on fossil fuels. Increasingly, scientists are in agreement that extensive climate change is occurring and is a result of human activities, especially the burning of fossil fuels. Even if discoveries of previously unknown fossil fuel reserves and technological breakthroughs were able to ensure a never-ending supply, these products could not be used without catastrophic climate implications. Concerns associated with climate change are discussed in more detail in Chapter Four.

Socioeconomic and Community Implications

The fossil fuel industry is one of our nation's most important employers. The coal industry alone provides more than one million jobs in the United States. Any transition away from fossil fuels will have significant consequences, especially in communities with energy-based economies. For example, Campbell County is located in the northeastern portion of Wyoming. Historically Campbell County has been economically dependent on ranching, and for decades it experienced a gradual economic and demographic decline. The 1930 census counted 6,720 residents in the county, but by 1960 this number had reduced to 5,861. However, Campbell County and the Powder River Basin then became the heart of the nation's coal development efforts. Numerous coal-mining jobs were created, quickly followed by construction jobs and employment in power plants. Between 1960 and 1970 the Campbell County population more than doubled to 12,957, and then nearly doubled again to 24,370 in 1980. The 2010 census lists the population of Campbell County at 46,133. Coal mining and electricity generation have transformed Campbell County and its primary city, Gillette, from a sleepy ranching community to the nation's most important coal-producing area.

Initially, rapid growth resulting from energy development during the 1970s led to the city of Gillette earning a reputation as a "boomtown." The term "Gillette Syndrome" was coined to describe social disruptions that occur when the population of a settlement grows too rapidly and the community is unable to respond adequately. Throughout the west during the 1970s and 1980s, communities were described by the extent to which they were experiencing Gillette Syndrome symptoms such as high crime rates, high rates of alcohol and drug abuse, degraded mental health and weakened social and community bonds (Brown et al. 2005; Kohrs 1974; Smith et al. 2001).

Gillette is no longer afflicted by boomtown-type growth. It is now a thriving community with good schools and good community services. Many existing jobs in the energy industry in Campbell County require significant skills and pay very well. In fact, in 2009 the median household income in Campbell County was higher than in any other nonmetropolitan county in the west with the exception of Los Alamos, New Mexico. This same process has also occurred in other western energy boomtowns (Smith and Krannich 2000), with communities dependent on the energy industry counting among the most prosperous in the west. However, questions remain about the long-term economic implications of an economy based on natural resource extraction and exportation. Many communities have experienced a "bust" when the resource was exhausted. Further, research has shown that economic development is often less extensive in communities with a resource-based economy compared to other communities (James and Aadland 2011; Kilkenny and Partridge 2009).

The emergence of fracking has resulted in a number of modern energy boomtowns (Christopherson and Rightor 2012). While these communities may eventually benefit from well-paying jobs and other sources of energy revenue, they are currently overwhelmed by a lack of housing and other infrastructure and are striving to cope with boom-like growth.

Major changes in the extent to which our nation is dependent on fossil fuel energy will obviously have major economic implications. Some employment skills will become obsolete and different ones will be needed. Also, there will be a transition in where jobs are located, which will have major implications for some communities. Thus, communities where alternative energy jobs are located may not be the same communities that host large numbers of jobs in the fossil fuel industries.

Reducing Fossil Fuel Dependence

As a consequence of problems associated with a fossil fuel economy, there is extensive interest in and effort toward decreasing our dependence on fossil fuels. Two major approaches for reducing fossil fuel use are increased alternative energy use and conservation.

Alternative Energy

Two major sources of non-fossil fuel energy include nuclear power and renewable energy. Nuclear power is generated by a nuclear fission process with reactors heating water to produce steam, which is used to power turbines to generate electricity. Nuclear power is discussed in Chapter Five.

The other major alternative energy sources are renewable energy. As opposed to nonrenewable fossil fuels, renewable energy is generated from naturally replenishing resources (IPCC 2011). Historically, the five major sources of renewable energy have been biomass, water, wind, geothermal and solar. While biomass has the potential to contribute to transportation fuel supplies, other renewable energy sources primarily provide electricity generation to reduce the amount of coal that is used. Table 1.7 presents data on renewable energy production from 1950 to 2011. In 2011, 83 percent of the renewable energy produced in the United States came from biomass (49 percent) and water (mostly hydroelectric; 34 percent). In recent years renewable energy production has increased substantially, especially wind and solar. At present, the primary factor preventing greater use of renewable energy is price. Energy from fossil fuels is simply cheaper to produce than renewable energy, especially when fossil fuel producers are not required to pay external costs such as those related to the production of greenhouse gasses and other pollutants. Cap and trade and other policy options could greatly increase the economic competitiveness of renewable energy by requiring fossil fuels to pay for the damage they create. Biofuel energy is discussed in Chapter Six, while wind and solar energy sources are the topic of Chapter Seven.

Conservation and Efficiency

An absolute necessity to assure a reasonable energy future is improved conservation. Vast amounts of energy could be saved if cars achieved better gas mileage and were driven fewer miles, if public transportation was more widely used, if we all walked more, and if homes were smaller and better insulated. Table 1.8 shows evidence of some steady improvements in energy efficiency. The amount of energy consumed per person has been fairly steady since about 1970 with a slight decline in recent years. Even though we

Table 1.7 Renewable Energy Production by Type in the United States, 1950–2011 (Quadrillion BTU)

Year	*Hydroelectric Power*	*Geothermal*	*Solar*	*Wind*	*Biomass*
1950	1.415	–	–	–	1.562
1955	1.360	–	–	–	1.424
1960	1.608	0.001	–	–	1.320
1965	2.059	0.002	–	–	1.335
1970	2.634	0.006	–	–	1.431
1975	3.155	0.034	–	–	1.499
1980	2.900	0.053	–	–	2.476
1985	2.970	0.097	–	–	3.016
1990	3.046	0.171	0.060	0.059	2.735
1995	3.205	0.152	0.070	0.069	3.099
2000	2.811	0.164	0.066	0.066	3.006
2005	2.703	0.181	0.063	0.178	3.104
2011	3.171	0.226	0.158	1.168	4.511

Source: Energy Information Administration, U.S. Department of Energy.

are, living in larger homes, driving more miles and using more electrical gadgets we are actually consuming less energy because of improvements in efficiency. Additionally, energy consumption per real dollars of GDP has steadily declined, indicating better energy efficiency in all segments of American life. Obviously, however, there is still much room for improvement.

Table 1.8 Energy Consumption per Person and per Real Dollars of GDP in the United States, 1950–2008

Year	*Energy Consumption per Person (Million BTU)*	*Energy Consumption per Real Dollars of GDP**
1950	227	19.48
1955	242	18.17
1960	250	18.02
1965	278	16.93
1970	331	17.99
1975	333	16.70
1980	344	15.13
1985	322	12.64
1990	339	11.90
1995	342	11.35
2000	351	10.08
2005	340	9.14
2008	327	8.52

*Thousand BTU per claimed in 2000 Dollars

Source: Energy Information Administration, U.S. Department of Energy.

Table 1.9 Motor Vehicle Mileage and Fuel Rates in the United States, 1950–2007

	Passenger Cars		*Vans, Pickup Trucks & SUVs*		*Trucks*		*All Motor Vehicles*	
Year	*Miles per Vehicle*	*Miles per Gallon*	*Miles per Vehicle*	*Miles per Gallon*	*Miles per Vehicle*	*Miles per Gallon*	*Miles per Vehicle*	*Miles per Gallon*
1950	9,060	15.0	–	–	10,316	8.4	9,321	12.8
1955	9,447	14.6	–	–	10,576	8.2	9,661	12.7
1960	9,518	14.3	–	–	10,693	8.0	9,732	12.4
1965	9,603	14.5	–	–	10,851	7.8	9,826	12.5
1970	9,989	13.5	8,676	10.0	13,565	5.5	9,978	12.0
1975	9,309	14.0	9,829	10.5	15,167	5.6	9,627	12.2
1980	8,813	16.0	10,437	12.2	18,736	5.4	9,458	13.3
1985	9,419	17.5	10,506	14.3	20,597	5.8	10,020	14.6
1990	10,504	20.2	11,902	16.1	23,603	6.0	11,107	16.4
1995	11,203	21.1	12,018	17.3	26,514	6.1	11,793	16.8
2000	11,976	21.9	11,672	17.4	25,617	5.8	12,164	16.9
2005	12,510	22.1	10,920	17.7	26,235	6.0	12,082	17.1
2007	12,293	22.5	10,952	18.0	25,141	5.9	11,910	17.2

Source: Energy Information Administration, U.S. Department of Energy.

A sector with the potential for vast improvements is transportation. Energy efficiency improvements in transportation are especially vital because transportation requires liquid fuels which cannot be offset by nuclear, wind, geothermal, solar or other alternative or renewable energy sources (unless electrical cars become more widely used). Table 1.9 shows motor vehicle mileage and fuel rates in the United States from 1950 to 2007. This table shows that the number of miles driven per vehicle steadily increased, until a slight decline which began in 2005. Also, since 1970 there has been a slow but steady increase in miles per gallon. Certainly there is the capacity for vast improvements. Without harming national productivity, it is possible to significantly reduce the number of miles driven and to improve the fuel rates of our vehicles. Many of our cities are not constructed in ways that are conducive to mass transit, and car-pooling rates could be improved. Chapter Eight discusses conservation and efficiency in more depth.

Conclusions

The development and use of fossil fuels has had dramatic consequences. At one level, by greatly increasing the capacity of farm and factory workers, our nation's GDP and our personal incomes have grown. By producing food and other necessities of life so efficiently, fossil fuels release many people to be artists, academics and athletes, which greatly add to our quality of life. However, a fossil fuel economy is unsustainable. Finite resource supplies are being depleted and the global climate implications of fossil fuel use are potentially catastrophic. Thus, efforts to enhance the production of alternative and renewable energy are vital, and time is of the essence. The ultimate advantage of the sun and wind, compared to fossil fuels, is that these energy sources are truly infinite.

References

Brown, R.B., S.F. Dorius, and R.S. Krannich. 2005. "The Boom-Bust-Recovery Cycle: Dynamics and Change in Community Satisfaction and Social Integration in Delta, Utah." *Rural Sociology* 70: 28–49.

Christopherson, Susan and Ned Rightor. 2012. "How Shale Gas Extraction Affects Drilling Localities: Lessons for Regional and City Policy Makers." *Journal of Town and City Management* Volume 2, Number 4.

Energy Information Administration. 2013. Washington, DC: U.S. Department of Energy.

Geri, Laurance R. and David E. McNabb. 2011. *Energy Policy in the U.S.* Boca Raton: CRC Press.

International Energy Agency. 2011. *World Energy Outlook*. Paris, France: OECD/IEA.

IPCC (Intergovernmental Panel on Climate Change). 2011. *IPCC Special Report on Renewable Energy Sources and Climate Change Mitigation.* New York: Cambridge University Press.

James, Alex and David Aadland. 2011. "The Curse of Natural Resources: An Empirical Investigation of U.S. Counties." *Resource and Energy Economics* 33(2): 440–453.
Jaramillo, P., C. Samaras, H. Wakely, and K. Meisterling. 2009. "Greenhouse Gas Implications of Using Coal for Transportation: Life-Cycle Assessment of Coal-to-Liquids, Plug-in Hybrids, and Hydrogen Pathways." *Energy Policy:* Elsevier.
Kilkenny, Maureen and Mark D. Partridge. 2009. "Export Sectors and Rural Development." *American Journal of Agricultural Economics* 91(4): 910–929.
Kohrs, El Dean. 1974. "Social Consequences of Boom Growth in Wyoming." Paper presented at the Annual Meeting of the Rocky Mountain Section of the American Association for the Advancement of Science, Laramie, Wyoming.
Lester, Richard K. and David M. Hart. 2012. *Unlocking Energy Innovation.* Cambridge, MA: The MIT Press.
McKibben, Bill. 2007. *Deep Economy.* New York: Times Books.
Sioshansi, F. 2011. *Energy, Sustainability, and the Environment: Technology, Incentives, Behavior.* Amsterdam: Elsevier.
Smith, Michael D. and Richard S. Krannich. 2000. "'Culture Clash' Revisited: Newcomers and Longer-Term Residents' Attitudes Toward Land Use, Development, and Environmental Issues in Rural Communities in the Rocky Mountain West." *Rural Sociology* 65(3): 396–421.
Smith, M.D., R.S. Krannich, and L.M. Hunter. 2001. "Growth, Decline, Stability and Disruption: A Longitudinal Analysis of Social Well-Being in Four Western Rural Communities." *Rural Sociology* 66(3): 425–456.
Stiller, David. 2000. *Wounding the West.* Lincoln: University of Nebraska Press.
Wilshire, Howard G., Jane E. Nielson, and Richard W. Hazlett. 2008. *The American West at Risk.* New York: Oxford University Press.
Yergin, Daniel. 2011. *The Quest.* New York: The Penguin Press.

2 Economic Implications of Unconventional Fossil Fuel Production

Amanda Weinstein and Mark Partridge

Introduction

Beginning around 2005, oil and gas markets in the U.S. took an unexpected turn. For decades domestic oil and gas production had been unable to meet demand, and imports rose despite national calls for energy security and independence. Innovations in oil and gas extraction, specifically hydraulic fracturing, dramatically reversed these trends. Hydraulic fracturing opened up oil and gas reserves in previously uneconomical shale plays across the U.S. and the world; "play" refers to a particular energy reserve under consideration for drilling. Oil and gas production from shale plays around the U.S. have been steadily rising as hydraulic fracturing becomes more widespread. In 2011 shale gas production accounted for 30 percent of natural gas production in the U.S., and tight oil production (oil extracted from shale plays) accounted for 33 percent of onshore oil production in the continental U.S. (EIA). As this trend continues, the U.S. is expected to become the world's largest natural gas producer in 2015 and the world's largest oil producer by 2017 (International Energy Agency [IEA], 2012).

Hydraulic fracturing is bringing about dramatic changes in U.S. energy markets. It also brings changes directly to the communities that happen to be located atop shale resources. This chapter lays out some of the expected and unexpected economic effects of shale development on local regions. We first outline characteristics of the national energy and gas industry and then trace the direct impacts in regions experiencing drilling. Our particular focus will be on the drilling impacts in Pennsylvania and North Dakota. We then describe other offsetting effects on the economy by describing how past energy booms have affected national and local economies. We conclude by describing how affected communities can avoid the boom–bust energy cycle that reduces economic well-being, as well as avoid the "natural resources curse" where energy development leads to lower long-term economic growth (Van der Ploeg, 2011). We describe the need for proper benefit–cost analysis instead of simply counting jobs in determining the best policy responses.

Hydraulic Fracturing

We start with a brief overview of the hydraulic fracturing method of extracting oil and gas resources, as it provides insight into the expected impacts. Before any drilling activity begins, a site must be selected and prepared. This includes establishing leasing agreements with the landowners. Oil and gas companies typically pay a per acre fee to lease the land in addition to royalty payments for any producing wells. For many landowners, the additional income provided by shale development is a significant windfall. This injection of income into an area may help boost the local economy, if only temporarily. However, Kelsey et al. (2011) find that approximately 55 percent of royalty money is saved rather than spent in the local economy. Additionally, royalty payments may be spent outside the local area (where better consumption amenities are located), especially in the case of absentee landowners, further reducing the local impacts of royalty payments.

Once a leasing agreement is in place, there is significant construction activity before the hydraulic fracturing process can begin. It takes between one and two months to prepare the site and construct the drill pad. Then there is about one month of rig work involving horizontal drilling. The hydraulic fracturing process itself takes between two to five days (Paleontological Research Institution, 2012). It involves a pressurized injection of one to eight million gallons of water that must be trucked onto the site. The pressure causes the shale to crack or fracture. About 2 percent of the water mixture is composed of chemical additives and sand (or other forms of proppants) which keeps the fissures open, allowing the trapped oil and gas resources to escape. About half of the water (this can vary substantially by site) returns to the surface as produced water or wastewater and eventually must be trucked to a disposal site (such as an injection well) or recycled. Once the well is producing oil or gas it can continue to produce for up to thirty years, though flow rates are highest in the first few weeks after which there is significant tapering over time. At this point the amount of oil and gas employment associated with the well is minimal.

The various stages of shale development have different employment and earnings impacts on the local area. Basic construction and trucking needs in the first stages may be met by local contractors. However, in areas without an existing drilling infrastructure, many of the highly-skilled drilling crews will likely come from out of state (Kelsey et al., 2011). Local area workers may require training in order to qualify for these relatively well-paid oil and gas jobs. The influx of oil and gas workers will increase the demand for various services such as restaurants, bars, and hotels. However, once the well has been fractured, its direct and indirect employment impact will decline significantly, especially once the supportive infrastructure such as pipelines is completed.

Shale Development across the U.S.

The Barnett shale in Texas was the first major shale play with the commercial production of natural gas from shale. Innovations in hydraulic fracturing and horizontal drilling that were proven in Texas opened up shale plays across the U.S. for development which had previously been deemed uneconomical. A map from the U.S. Energy Information Administration (EIA) shows the various shale plays in the U.S. (Figure 2.1).

Developments made to the hydraulic fracturing process in Texas have brought shale development into new areas such as the Marcellus shale in Pennsylvania. The EIA indicates that the Marcellus shale play is the largest shale play in terms of recoverable natural gas, containing over half of the total recoverable natural gas in the U.S. (EIA, June 2012). The Haynesville and Barnett plays are second and third, respectively. The largest shale plays in terms of recoverable oil according to the EIA are the Monterey, Bakken, and Eagle Ford plays. Figures 2.2 and 2.3 show the production of oil and gas by shale play. Shale oil and gas production across the U.S. has been steadily increasing since about 2005 and is expected to continue to rise.

A boom in the number of oil and gas jobs has accompanied the boom in production (Figure 2.4). A temporary decrease in oil and gas employment is evident in 2008 because of a sudden price drop in oil and a drop in natural

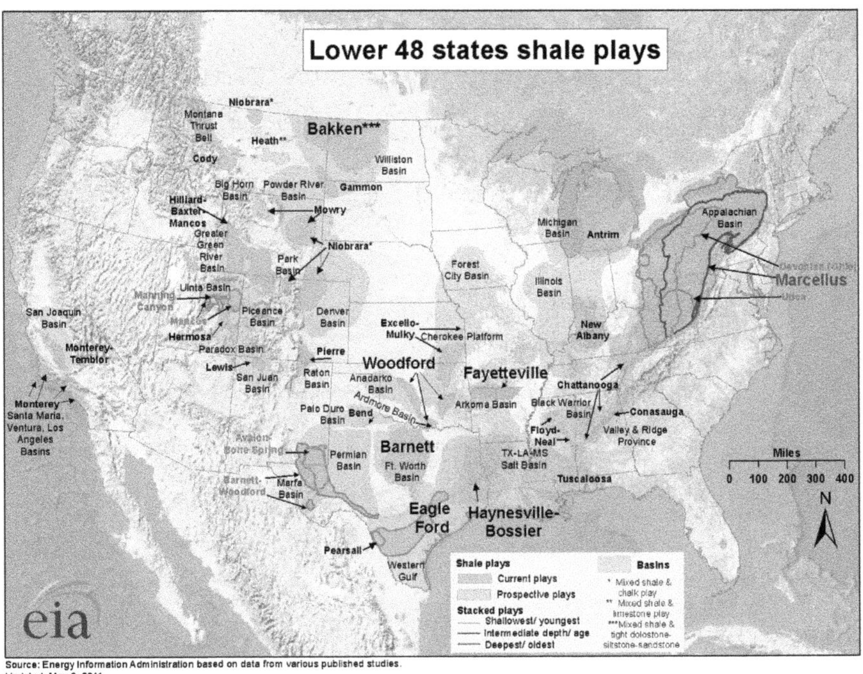

Figure 2.1 U.S. Shale Plays.
Source: EIA.

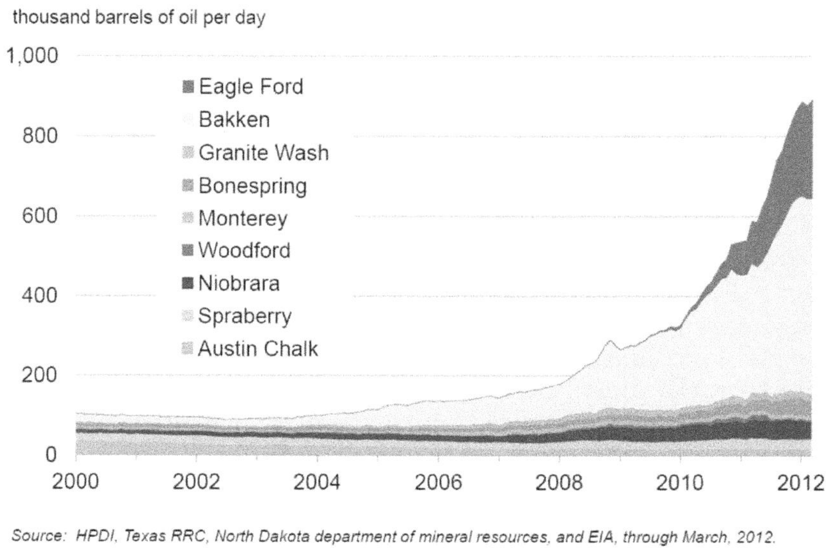

Figure 2.2 Oil Production from Shale.
Source: EIA (June, 2012) www.eia.gov/pressroom/presentations/staub_06272012.pdf.

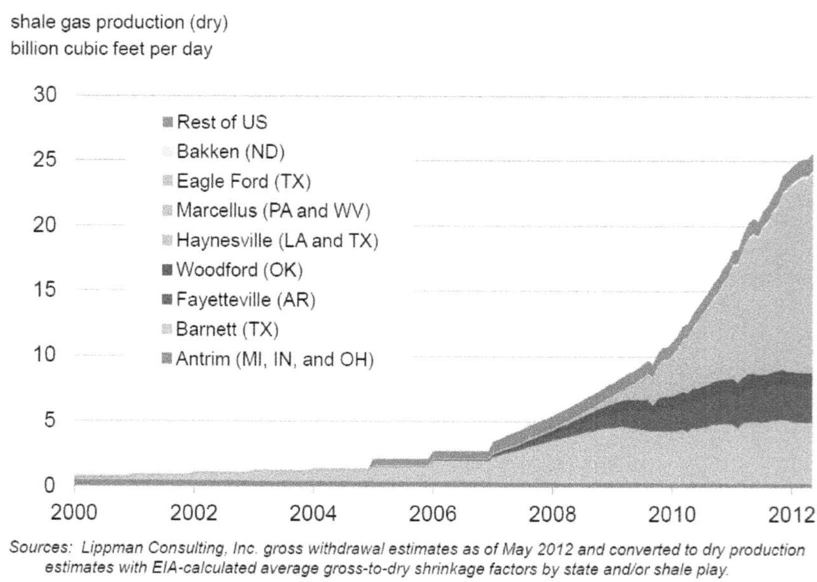

Figure 2.3 Natural Gas Production from Shale.
Source: EIA (June, 2012) www.eia.gov/pressroom/presentations/staub_06272012.pdf.

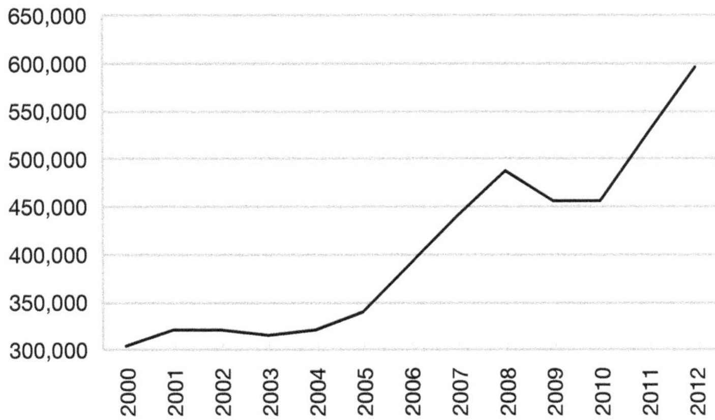

Figure 2.4 Oil and Gas Employment Boom.
Source: BLS CES http://data.bls.gov/pdq/querytool.jsp?survey=ce NAICS Industry codes: 21111 – Oil and Gas Extraction, 213112 – Support Activities for Oil and Gas Operations, 237120 – Oil and Gas Pipeline Construction.

gas prices due in part to the glut of shale production. Since the drop, oil prices have been steadily increasing but natural gas prices have remained low. Lower energy prices are generally beneficial for residents and firms across the U.S. However, the dramatic rise in shale development and innovations in oil and gas extraction itself have been fueled by high and increasing energy prices that date back to the late 1990s. A moderation in energy prices would slow further development, though oil and gas prices are expected to rise in the future (*The Economist*, 2012).

The employment effects of shale development are not uniform across the U.S. Shale development (along with oil and gas job growth) has been focused on those areas with the most productive shale plays (typically the deeper shale plays). Oil and gas employment growth for those states most affected by shale development is shown in Figure 2.5, benchmarked to 100 at the year 2005 (excluding North Dakota, for reasons described below). Ohio provides a good counterfactual to compare with other states as shale drilling did not begin there in earnest until 2012. Shale drilling states appear to have experienced a significant boost in direct employment in the oil and natural gas industry.

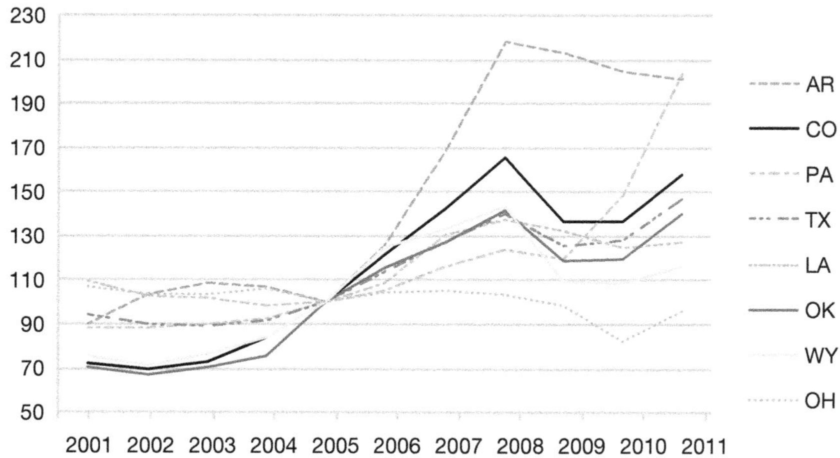

Figure 2.5 State Oil and Gas Employment Growth.
Source: BLS Quarterly Census of Employment and Wages. The respective NAICS codes are: 21111 – Oil and Gas Extraction, 213111 – Drilling Oil and Gas Wells, 213112 – Support Activities for Oil and Gas Operations, 541360 – Geophysical Surveying and Mapping Services, 238912 – Nonresidential Site Preparation Contractors, 333132 – Oil and Gas Field Machinery and Equipment Manufacturing, 486210 – Pipeline Transportation of Natural Gas, 237120 – Oil and Gas Pipeline Construction.

North Dakota's Shale Boom

It is important to get a better understanding of the regional economic benefits of shale development as many of the costs of shale development are localized. We begin by examining the economic benefits of the shale boom in North Dakota. Because of its oil resources, the Bakken shale in northwestern North Dakota has experienced the most pronounced boom in production and employment growth. Figure 2.6 shows the dramatic increase in North Dakota's oil and gas employment growth (benchmarked at 100 for the year 2005). Its mining employment (including coal and other mining) went from 3,600 in 2003 to 24,600 in March 2013 (U.S. Department of Labor, B.L.S). However, even though it shows impressive growth rates, the mining industry is still just a small share (under 6 percent) of North Dakota's relatively small total non-agricultural employment of over 429,800 (BLS, CES, March 2013). Oil and gas extraction is a capital-intensive industry; thus, output is associated with fewer workers per dollar of output than is typical in the wider U.S. economy. Overall, using U.S. Bureau of Labor Statistics data from the *Quarterly Census of Employment and Wages*, the entire Bakken region in North Dakota and Montana experienced an increase in jobs of around 49,000 from 2003 to mid-2012, although it is not clear if all of these are due to energy development. For perspective, the entire U.S. economy has averaged an increase of about 160,000 jobs a month since it began expanding in February 2010.

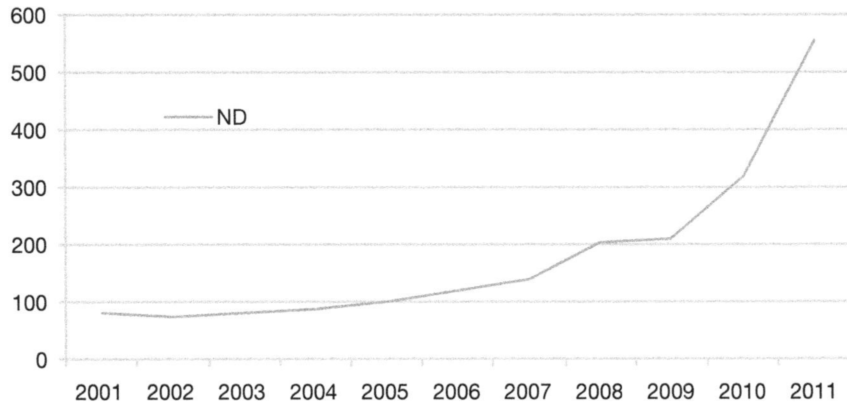

Figure 2.6 North Dakota Oil and Gas Employment Growth.
Source: BLS Quarterly Census of Employment and Wages statewide estimates for North Dakota. NAICs codes are: 21111 – Oil and Gas Extraction, 213111 – Drilling Oil and Gas Wells, 213112 – Support Activities for Oil and Gas Operations, 541360 – Geophysical Surveying and Mapping Services, 238912 – Nonresidential Site Preparation Contractors, 333132 – Oil and Gas Field Machinery and Equipment Manufacturing, 486210 – Pipeline Transportation of Natural Gas, 237120 – Oil and Gas Pipeline Construction.

Although mining employment remains a small share of the overall economy of North Dakota, it may still have a large impact on the particularly rural and remote areas of northwestern North Dakota. Consider the case of the small town of Williston in Williams County, ND. Williams County, ND is located in the middle of the Bakken shale and has experienced dramatic shale development (the most within North Dakota and the U.S.). Total employment in Williams County remained just under 13,000 until 2005, at which point employment began to climb, reaching just over 28,000 in 2011 (BEA).

More than doubling the employment of a rural area such as Williston can have significant impacts on the local area and the local economy. This employment increase is not entirely from oil and gas workers. The influx of oil and gas workers into the area increases the demand for various goods and services such as oil industry suppliers, restaurants, bars, and hotels. The increased demand for labor raises employment as well as wages. The increase in wages, along with the relatively well-paid oil and gas jobs, increases income in the area. From 2005 to 2011 the average per capita income in Williams County increased by 166 percent and was over $81,000 in 2011 (BEA). However, the increase in real *wages* is smaller than the nominal increase suggests because prices throughout the region also increased (and much of the *income* increase was in the form of royalties and lease payments to landowners). The increase in demand for housing and other goods has pushed up prices throughout the region, increasing rent from an average of $500 per month in 2005 to approximately $2,000 (Shactman 2012). Prices for gasoline and groceries are 30 percent higher (Oldham, Jan 2012).

Williston will also benefit from an increase in tax revenue from drilling activities. For North Dakota as a whole, oil companies pay approximately 11.5 percent in taxes which amounted to approximately $2.6 billion in 2008 (Oldham, Jan 2012 and Feb 2012). This new revenue has helped fund projects to improve drinking water supplies and update the state prison in Bismarck, but the local share of oil and gas taxes is only 11 percent (Headwaters Economics, 2012). Of the $1.2 billion that has been set aside so far to help drilling counties, about $885 million remains to be distributed. Many in Williston and other drilling areas say it is still not enough to cover the strains on their small towns (Oldham, Jan 2012 and Feb 2012). One of the most visible signs of strain is on Williston's infrastructure, most notably its roads and traffic. Rural gravel roads previously used mainly by farmers now have eight hundred trucks traverse them in a single day (Oldham, Feb 2012). Additionally, the sheer number of people flocking to this small city has put a considerable strain on its infrastructure and services including the sewer system, water system, and energy grid, as well as its police and school system. Williston is struggling to maintain its infrastructure, services, and amenities, let alone to improve them. Residents are trading off temporary, high-paying jobs in return for a reduced quality of life. Williston officials hope this boom will last at least ten years, but they expect to retain only about 30 percent of the new workers afterwards. With limited public infrastructure, amenities stretched and employment waning, there may be few reasons for these families to stay when the boom ends.

The economic impacts felt by North Dakota and particularly Williston are felt more acutely because its *relative* development has been the most dramatic in the U.S. Shale development will most likely impact other areas much more moderately, especially given that northwestern North Dakota is among the most remote and sparsely populated regions in the country. However, as a case study Williston highlights the economic benefits and costs that accompany shale development, although a more rigorous analysis is necessary to estimate the expected economic impact of shale development.

The Expected Economic Value of Shale Development

Arguably, the most important impact of shale development to these communities is the jobs created by the industry. Many of these communities are located in rural regions that have long struggled. Initial estimates of the number of jobs that would be created from media reports and impact studies appear to have been exaggerated (Weinstein and Partridge, 2011). Inflated job estimates are often used to justify various financial incentives and regulatory changes as a way to encourage development across all industries. Thus, it is important to have an accurate estimate of the economic impact of shale development so that communities and their residents can make more accurate plans in terms of economic development and in weighing the environmental impacts.

Impact Studies

Impact studies typically utilize an input–output type methodology to estimate how many jobs are *supported* by an industry. They usually estimate three types of employment effects: (1) the jobs directly employed in oil and gas mining; (2) the jobs indirectly associated with oil and gas mining, such as pipeline construction; and (3) the jobs induced by the additional household income such as restaurant, bar, hotel, and convenience store workers (Weinstein and Partridge, 2011). The summation of the three types of employment effects is the total number of jobs *supported* by the industry. The multiplier estimate approximates how many direct, indirect and induced jobs are supported by a given economic activity. Impact studies commonly overestimate the net number of jobs created (or supported) by an industry, producing multiplier estimates that are not supported by ex post analyses, particularly when short-term construction projects are involved in the economic impact.

There are several reasons why impact studies often overestimate *net* economic impacts. Impact studies sometimes incorporate unrealistic assumptions that do not accurately reflect the economy. For example, the percentage of input spending and hiring that remains within the region of study can be overstated, as in current shale gas studies such as those by Considine et al. (2011) and Kleinhenz & Associates (2011) which use assumptions that 95 percent and 90 percent of the money remains in the region, respectively. These assumptions are unrealistic for a variety of reasons, including the fact that shale development is relatively new and undeveloped in many areas receiving the activity. Thus, more of the inputs would likely be brought in from outside the state, particularly for states that are not large historic oil and gas producers. Rather, many of these inputs might come from a state like Texas with a much deeper oil and gas supply chain and infrastructure. In terms of labor inputs, Kelsey et al. (2011) found that 37 percent of the Marcellus shale jobs have gone to out-of-state residents. Moreover, there is further leakage in spending as landowners save or invest approximately 55 percent of their royalties and lease payments (Kelsey et al., 2011). Finally, even local residents purchase many of their goods outside of the area (Kelsey et al., 2011).

Some impact studies also count indirect and induced jobs as direct jobs before applying the multiplier, thereby double-counting the economic effects and possibly vastly increasing estimates (e.g. see the discussion in Weinstein and Partridge, 2011). Traditional input–output based impact studies also fail to account for some of the negative impacts associated with drilling activity such as the strain on infrastructure and public services or any environmental damage that may result, which may reduce tourism and other activities that are more directly tied to local amenities. Oil and gas extraction may also displace other activities that would have occurred had no shale development occurred. For example, as more natural gas is used to produce energy, less coal will be needed and thus fewer coal mining jobs. Therefore, for example, at the national level, many of the new jobs created around natural gas extraction will be at least partially offset by losses in coal mining. Oil and gas extraction may also crowd out employment in other sectors as local area wages and land prices increase, making those sectors less competitive (e.g. especially traded sectors as in Dutch Disease). Industries that rely on low wages, such as manufacturing, may be particularly affected.

Even a well-constructed impact study can cause confusion in the interpretation of the results by the public, media, and policymakers. Impact studies do not estimate continuous employment numbers as the public and media often presume. As discussed above, there are jobs associated with pipeline construction, site preparation, trucking, drilling, and maintenance once the well is producing. All of these jobs are included in the estimate of the economic impact, but many of these jobs, such as pipeline construction, are temporary. The public is likely to be more interested in permanent employment and may view estimates of the number of jobs supported by an industry as permanently created jobs. Moreover, the public often views these estimates as the number of jobs *created* when impact studies can only produce the number of jobs *supported*, overlooking possible offsetting losses if high wages make local employers less competitive or businesses avoid locating in boom regions to avoid high input costs. Input–output oriented impact studies cannot estimate the *net* number of jobs created and thus can be misleading.

A number of recent impact studies have been released touting the significant economic benefits of shale development. These studies include estimates for Pennsylvania, Ohio and California. An industry-funded study by Considine et al. (2011) found that shale development supported 140,000 jobs in Pennsylvania and yielded $11.2 billion in value added in just one year (during 2010). Using actual data, Pennsylvania's total direct natural gas extraction employment was just under 26,000 in 2010 after adding approximately 10,000 direct shale development jobs between 2004 and 2010 (BLS; see Figure 2.7). Using the total economic impact of 140,000 from Considine et al. (2011) in combination with the direct natural gas jobs created would imply a multiplier of approximately fourteen (meaning that for every new direct shale job created between 2004 and 2010, thirteen jobs were created or supported in other industries). To show just how large this estimate is, recall that in ten years (from 2003 to 2013) North Dakota's entire mining employment increased by 21,000. During this same timeframe, North Dakota's total non-agricultural employment increased by 110,000 (BLS). Even assuming that all 110,000 additional jobs were created or supported by shale development (which is highly unlikely), this still implies a multiplier of only 5.2 (for every mining job created, 4.2 jobs in other industries were created). Kelsey et al. (2011) performed an impact study using IMPLAN and estimated the economic impact of shale development in Pennsylvania. Their estimate was more moderate at just over 23,000 jobs and more than $3.1 billion in spending (during 2009). Their multiplier was in the range of 2.0, which is the number that would be supported in well-done impact studies.

Various other impact studies (mainly industry-funded) have since followed suit in producing large and unrealistic estimates of the economic impacts from shale development in other states. An impact study by Kleinhenz & Associates (2011) found that the natural gas industry would create and support over 200,000 jobs in Ohio and $14 billion in spending in the first four years of drilling. A USC Global Energy Network (2013) impact study found that shale development of the Monterey shale in central California would create $24.6 billion in state and local tax revenue and a whopping 2.8 million jobs, which would cause unemployment in the state to fall. However, a comparison of these predictions with the actual economic impact experienced by North Dakota during what has been perhaps the most pronounced shale boom in the U.S. shows that the forecasts are unrealistic. North Dakota's unemployment, although low, has barely changed from 3.6 percent in 2003 to 3.2 percent in 2013. Additionally, even though North Dakota has experienced a dramatic increase in mining employment, the mining industry's share of total employment remains small and thus is unlikely to have a large impact on, for example, state unemployment rates, especially since the unemployment rate was already low before the boom. With total employment of over seventeen million in California, it is hard to see how shale development would have a large impact on its unemployment rates.

Because of the problems associated with input–output oriented impact studies, economists do not view this type of analysis as best practice (see Irwin et al., 2010 and Partridge and Rickman, 2010 for additional discussion of the weaknesses of impact studies). Best practice includes approaches that account for various displacement effects and derive more accurate estimates of the actual effects based on the matching and experimental literatures. These approaches use actual data to estimate economic impacts, not estimates based on the assumptions in the economic model. A counterfactual of what would have happened in these areas if no shale development had occurred must be developed to compare with what actually happened. Typical impact studies fail to create a clear counterfactual of what would have happened in other industries if there was no impact, instead producing something akin to a before and after estimate. Rather, econometric studies employing modern approaches such as difference-in-difference calculations using actual employment data are needed. We describe such approaches below.

The Economic Impact of Shale Development in Pennsylvania and Ohio

One feature that typically leads to errors in economic impact assessments is that they do not compare their findings with similar places that have experienced similar-sized economic shocks—i.e. the notion of external validity. For example, Pennsylvania provides a good estimate of the number of jobs its neighbor Ohio can expect in its first years of shale development. Rural Ohio and Pennsylvania have similar topographies and land use patterns for drilling, and their economies have experienced similar trends for decades. Of course, there are differences in the impacts of an energy boom, such as the possibility that Ohio would have more liquids including oil, but drilling in both states would require similar infrastructure investments to begin production (and hence there would be similar direct employment impacts). Ohio began drilling in 2012. Weinstein and Partridge (2011) use BLS employment data for Ohio and Pennsylvania to estimate the economic impact of shale development in Pennsylvania from 2004 to 2010 and the expected impact for Ohio in its first years of development. Figure 2.7 shows that Pennsylvania gained approximately 10,000 direct natural gas jobs from 2004 to 2010. Assuming a realistic multiplier of approximately two (perhaps an overstatement), this implies the total number of jobs created in Pennsylvania from shale development was approximately 20,000 during the first six years of development. Weinstein and Partridge (2011) then predict that the path of development in Ohio would follow a similar course given the similarities in the two states' drilling regions, approximating to about 20,000 jobs in the early years of Ohio's shale development (or nearly 10 percent of the economic benefit estimate produced by Kleinhenz & Associates, 2011).[1] Figure 2.7 has been updated to include data for 2011. Using that additional year of data, the total employment impact from 2004 to 2011 is over 38,000 using a realistic but still generous multiplier of two (see Stabler and Olfert, 2002 and Hughes, 2003 for discussion of multipliers). Using actual direct employment data from a similar shock to estimate the economic impact, as opposed to using the estimated outcome of an impact computer model, is another advantage of this approach.

Despite Pennsylvania's impressive growth rates in direct natural gas employment, its unemployment rates have not improved. Unemployment rates in Pennsylvania have actually increased from 5.4 percent in 2004 to 7.5 percent as of May, 2013 (BLS), illustrating that other macro factors are much more important than an energy boom in driving its dynamics. The natural gas extraction industry should not be expected to have a large impact on the unemployment rates of Pennsylvania because the natural gas industry only represents a small proportion of total employment. From 2004 to 2011, the employment share of oil and natural gas increased from approximately 0.29 percent to 0.64 percent. This small employment share is simply not large enough to have a significant effect on the statewide economy (and state unemployment rates). Nonetheless, relatively small employment in the oil and natural gas industry should be viewed as a positive feature as it is indicative of the sector's high rate of productivity and it helps to lower energy costs in general, though these effects are much more broadly felt than local impact estimates.

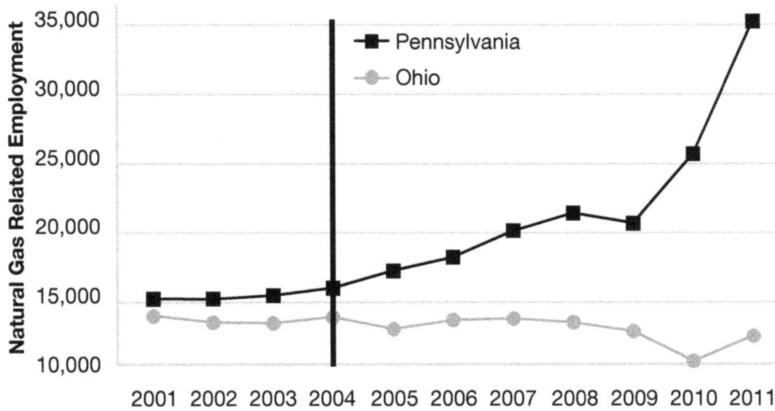

Figure 2.7 Ohio and Pennsylvania Natural Gas Employment.
Source: BLS http://data.bls.gov/pdq/querytool.jsp?survey=en (accessed May 2014). Updated chart from Weinstein and Partridge (2011).
Notes: The direct effects would commonly include the drilling and extraction activities while indirect effects would normally include inputs such as pipeline construction and field equipment manufacturing.

The Local Economic Benefit of Shale Development

Although shale development does not appear to have a dramatic effect on the statewide economy (and unemployment rates) in Pennsylvania, it may have had a more pronounced effect on the most impacted local areas, especially in rural counties where much of the shale development is taking place. Thus, Weinstein and Partridge (2011) took a closer look at the economic impact of shale development at the county level. Drilling counties in Pennsylvania tended to be more rural with lower employment growth before shale drilling began. Therefore, an analysis of all the non-drilling counties in Pennsylvania does not make a good counterfactual for the state's drilling counties. Weinstein and Partridge match selected drilling counties in the northeast and southwest regions of Pennsylvania with similar non-drilling counties to compare their economic outcomes before the shale drilling period (2001 to 2005) and during shale drilling (2005 to 2009). The findings indicate that drilling counties consistently perform better in terms of per capita income growth in the first few years after drilling commences, but the employment benefits were only clear for those counties in the northeast region of Pennsylvania (although it is likely that employment benefits would be clearer for other counties when adding in data after 2010).

Weinstein and Partridge (2011) also utilize a more formal statistical approach, a difference-in-difference methodology, to estimate the economic impact of shale development on counties within Pennsylvania. County level employment and earnings growth are strongly related to inherent differences in counties, such as their quality of local government, distance to urban centers, and the educational and demographic attributes of the population. These effects are generally considered to be county fixed effects C_i. Equations 1 and 2 show the impact of the number of wells on the percentage employment growth (Y_{ij}) for county *i* in period *j* (where *j*=0 is the period before drilling and *j*=1 is the period during shale development). However, many of the county fixed effects (C_i) are unobserved, or else they are not amenable to measurement. Differencing Equation 2 and Equation 1 eliminates the need to account for these county fixed effects, and omitted variable bias should not be an issue (Equation 3). The effects of shale development in our example were measured by the local change in the number of wells drilled (or in other settings, the local change in employment in the energy industry).

$$Y_{i0}= \beta_0 + \beta_1(\text{Number of Wells})_{i0} + C_i + \varepsilon_{i0} \quad (1)$$
$$Y_{i1}= \beta_0 + \beta_1(\text{Number of Wells})_{i1} + C_i + \varepsilon_{i1} \quad (2)$$
$$Y_{i1}- Y_{i0}=\beta_0 + \beta_1(\Delta \text{ Number of Wells}) + \varepsilon_i \quad (3)$$

An empirical estimation of Equation 3 for all counties in Pennsylvania shows that drilling has a small and statistically non-significant impact on employment growth for drilling counties, but a clear positive and significant impact on per capita income growth (Weinstein and Partridge, 2011). Using a similar difference-in-difference methodology, Weber (2012) finds that every $1 million in shale gas production is associated with 2.35 jobs within counties in Texas, Colorado, and Wyoming.

Weinstein (2014) uses a variant of the difference-in-difference methodology to create a counterfactual for U.S. counties in the lower forty-eight states (Equation 4). This equation controls for the difference between drilling and non-drilling counties (β_2) to estimate the impact of shale development (β_3). Using local oil and gas employment to measure shale development in drilling counties across the U.S., Weinstein (2014) finds that for every oil and gas job created, approximately 0.3 jobs are created on *net* in other industries (mainly the non-traded goods sector). This is a multiplier of 1.3 (far less than the multipliers estimated by impact studies that rely on input–output analysis). Weinstein finds that the impact on earnings is nearly double the impact on employment for counties.

$$Y_{it} = \beta_0 + \beta_1(\text{Drilling Period}) + \beta_2(\text{Drilling County}) + \beta_3(\text{Drilling Period*Drilling County}) + \varepsilon_i \quad (4)$$

Using carefully-designed statistical methodologies to create counterfactuals produces more moderate estimates of the net impact of shale development on local area employment. The larger relative net impact on earnings compared to employment is most likely due to the higher wages associated with the oil and gas industries, in addition to leasing and royalty payments.

Previous Energy Booms and Long Run Economic Impacts

The long-term economic implications of natural resource extraction are often ignored in economic impact studies. There is hope that a natural resource boom will provide sufficient agglomeration economies for economic growth to take off (Michaels, 2010). However, Rosenthal and Strange (2001) find that natural resources used for energy have no significant agglomeration effects. Williston, ND provides a good case study because it experienced an energy boom in the late 1970s and early 1980s, when oil spiked, and a bust when prices fell thereafter (Figure 2.8). Williston's economy subsequently lagged significantly behind the overall U.S. economy up until the most recent shale boom. Williston (Williams County) did not surpass its 1981 peak until 2010 (U.S. BEA).

There is a rich history of energy booms in the U.S. and they are often followed by busts, as seen in the 1980s. Black et al. (2005) found that the negative economic impacts of the coal bust of the 1980s were larger than the positive impacts of the coal boom that preceded it in the 1970s. They found that 3.5 jobs were lost in other industries for every ten coal jobs lost, compared with the coal boom when two jobs were created for every ten coal jobs created, implying a multiplier of 1.2.

The relatively variable long-run economic performance of Williston is not an aberration. There seem to be more examples of underperforming energy

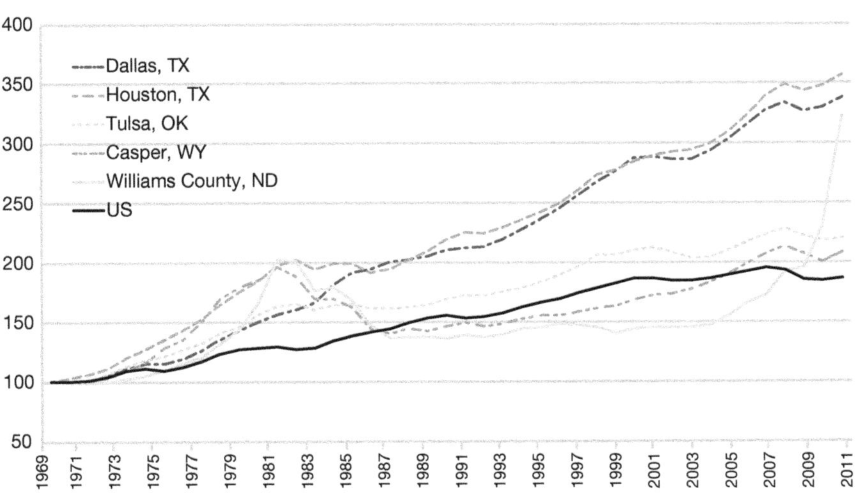

Figure 2.8 Employment Growth over Time.
Source: BEA (reproduced and updated from Farren et al., 2012).

economies in places such as West Virginia, Louisiana, Nigeria, and Venezuela than there are of over-performing energy economies like Houston, Dallas, and Alberta, Canada. Figure 2.8 is suggestive that better long-term economic performance in natural resource-led economies tends to occur in larger cities that contain the headquarters and higher-order functioning of the industry, such as Houston. Poor long-term economic performance is common in extractive resource-based economies at the source of the extraction (which is typically rural).

This natural resource curse has been documented at every level of geography, from countries (Sachs and Warner, 1995) to U.S. states and counties (Papyrakis and Gerlagh, 2007; Kilkenny and Partridge, 2009; James and Aadland, 2011). Kraybill and Dorfman's (1992) results suggest that the long run cumulative multiplier of the natural resource sector is even negative using a computable general equilibrium model (see Van der Ploeg, 2011 and Van der Ploeg and Venables, 2012 for reviews of the literature).

Various reasons for the poor performance of energy economies have been suggested. As mentioned previously, energy development can displace other economic activity (i.e. Dutch Disease). Increasing labor and land costs in areas with energy development decrease an area's competitiveness, which can have long-term implications especially in the tradable sectors such as manufacturing (Black et al., 2005). Additionally, with relatively high paying low skilled jobs readily available, there is less incentive for residents to invest in education and general skill level (Gylfason, 2001). Lower human capital levels of resource intensive areas can reduce long-term innovation, productivity, and economic growth (Benhabib and Spiegel, 1994; Glaeser et al., 1995; Simon, 1998). An increased dependence on the natural resources sector can also lead to a less diverse economy. Lower industry diversity is associated with higher volatility as the area becomes more vulnerable to economic shocks (Hammond and Thompson, 2004; Gunton, 2003; Randall and Ironside, 1996; Izraeli and Murphy 2003). Moreover, regions that are dependent on energy producing natural resources will also be particularly affected by volatile energy prices. Kraybill and Dorfman (1992) found that a shock in the natural resource sector causes more instability and will take longer to work through the economy than shocks in other sectors (e.g. see Partridge and Rickman, 2003 and Van der Ploeg, 2011). There may be significant economic consequences as resource dependent areas experience a depreciation of natural amenities and other negative environmental externalities associated with natural resource extraction. Finally, government institutions can exacerbate the negative long-run economic effects of natural resource extraction (Rodriguez and Sachs, 1999). For example, the increased spending may cause corruption and rent-seeking, or the local government may simply become too accustomed to a level of spending that can't be sustained. Good government policies are essential to help mitigate the negative effects that are possible with shale development and potentially avoid the resource curse over the long run.

Policy Implications

It is important for regions to have realistic expectations of the short-term and long-term economic benefits of shale development. It is also important for regions to maximize the potential economic benefits from shale development by understanding and mitigating the effects of factors that contribute to the natural resource curse. Specialization in one industry, especially one that is closely linked to volatile energy prices, leaves an economy more vulnerable to shocks. Thus, efforts to diversify the economy are crucial for long-run growth. A narrow focus on shale development while ignoring other sectors, such as the service sector, can contribute to anemic growth in resource dependent areas. Because shale energy is an exhaustible resource, shale development involves the permanent depletion of natural capital. The Solow-Hartwick Rule states that it is important to offset the loss in natural capital through investment in either human capital or public capital, as well as counteract the environmental consequences and offset the degradation on infrastructure and public services (Hartwick, 1977). Investment in human capital will also help to boost education levels and counteract the disincentive for skill development provided by the relatively highly paid but low-skilled jobs in the oil and gas industry. Hence, industry taxes must be set appropriately to account for both the short-term and long-term costs of shale development. Local government institutions are critical to ensure funding is spent appropriately, and that the government doesn't become accustomed to an unsustainably high level of spending. By appropriately taxing the natural resource extraction industry and using the revenue to offset short-term and long-term costs, regions like Alberta, Canada and Dallas, TX have seemingly managed to avoid the resource curse (Farren et al., 2012).

Conclusion

The short-term impacts of energy development often include increased employment, although the largest impact appears to be on local incomes of select groups, most notably landlords. By creating a true counterfactual, the employment impacts of shale development can be more accurately estimated and seem to be much more moderate than initial impact studies and media reports have suggested. A more precise estimate of the economic benefits of shale development is a critically important piece in a larger Benefit Cost Analysis (BCA) of the value of shale development. The true employment and earnings benefits should be weighed against the various environmental and local costs of shale oil and gas extraction. However, any BCA of shale development should be careful to consider the larger energy implications in the analysis. For example, the benefits and costs of shale development should be compared to its closest alternative (which is coal in the case of shale natural gas extraction).

Moreover, an accurate estimate of the short- and long-term economic impacts of shale development is essential for a community to manage its economic future. In particular, communities should take steps to mitigate the long-term effects associated with the resource curse and ensure that they benefit from energy development in the long term. These steps should include taxing the industry to account for negative spillovers, diversifying their economies, building enhanced infrastructure, and investing in education and training. Likewise, local communities need to ensure the environment is adequately protected in order to promote long-term sustainable growth after the boom ends. This implies sensible regulation to protect a region's natural capital.

Note

1 Weinstein (2014) finds that the employment multiplier for shale development may actually be closer to 1.3.

References

Benhabib, Jess and Mark M. Spiegel. (1994) "The Role of Human Capital in Economic Development: Evidence from Aggregate Cross-County Data." *Journal of Monetary Economics*, Vol. 34, No. 2, pp. 143–173.

Black, Dan, Terra McKinnish, and Seth Sanders. (2005) "The Economic Impact of the Coal Boom and Bust." *The Economic Journal,* Vol. 115 No. 503, pp. 449–476.

Considine, Timothy, Robert Watson, and Seth Blumsack. (2011) "The Pennsylvania Marcellus Natural Gas Industry: Status, Economic Impacts and Future Potential." http://marcelluscoalition.org/wp-content/uploads/2011/07/Final-2011–PA-Marcellus-Economic-Impacts.pdf (accessed 20 October 2013).

The Economist. (2012) "Shale of the Century." www.economist.com/node/21556242 (accessed 20 October 2013).

Ellis, Blake. (2011) "When a Boomtown goes Bust: 'Sudden Desertion.'" CNNMoney. http://money.cnn.com/2011/12/06/pf/oil_boom_bust/index.htm (accessed 20 October 2013).

Farren, Michael, Amanda Weinstein, and Mark Partridge. (2012) "Making Shale Development Work for Ohio." The Swank Program in Rural-Urban Policy Summary and Report.

Glaeser, Edward L., José A. Scheinkman, and Andrei Shleifer. (1995) "Economic Growth in a Cross-Section of Cities." *Journal of Monetary Economics*, Vol. 36, No. 1, pp. 117–143.

Gunton, Thomas. (2003) "Natural Resources and Regional Development: An Assessment of Dependency and Comparative Advantage Paradigms." *Economic Geography*, Vol. 79, No. 1, pp. 67–94.

Gylfason, Thorvaldur. (2001) "Natural Resources, Education, and Economic Development." *European Economic Review* Vol. 45, No. 4–6, pp. 847–849.

Hammond, George W. and Eric Thompson. (2004) "Employment Risk in U.S. Metropolitan and Nonmetropolitan Regions: The Influence of Industrial

Specialization and Population Characteristics." *Journal of Regional Science*, Vol. 44, No. 2, pp. 517–542.

Hartwick, John M. (1977) "Intergenerational Equity and the Investment of Rents from Exhaustible Re-sources." *American Economic Review*, 67, December, pp. 972–974.

Headwaters Economics. "Benefiting from Unconventional Oil: State Fiscal Policy is Unprepared for the Heightened Community Impacts of Unconventional Oil Plays." Stanford University. The Bill Lane Center for the American West. http://headwaterseconomics.org/wphw/wp-content/uploads/ND_Unconventional_Oil_Communities.pdf (accessed 20 October 2013).

Hughes, David W. (2003) Policy Uses of Economic Multiplier and Impact Analysis. *Choices* 2nd Quarter. www.choicesmagazine.org/2003–2/2003–2–06.htm (accessed 20 October 2013).

International Energy Agency. (2012) "World Energy Outlook 2012." www.iea.org/publications/freepublications/publication/English.pdf (accessed 20 October 2013).

Irwin, Elena G., Andrew M. Isserman, Maureen Kilkenny, and Mark D. Partridge. (2010) "A Century of Research on Rural Development and Regional Issues." *American Journal of Agricultural Economics* 92(2): 522–553; doi: 10.1093/ajae/aaq008

Izraeli, Oded and Kevin J. Murphy. (2003) "The Effect of Industrial Diversity on State Unemployment Rate and Per Capita Income." *The Annals of Regional Science*, Vol. 37, pp. 1–14.

James, Alex and David Aadland. (2011). "The Curse of Natural Resources: An Empirical Investigation of U.S. Counties." *Resource and Energy Economics*, Vol. 33, No. 2, pp. 440–453.

Kelsey, Timothy W., Martin Shields, James Ladlee, and Melissa Ward. (2011). "Economic Impacts of Marcellus Shale in Pennsylvania: Employment and Income in 2009." Marcellus Shale Education & Training Center. www.marcellus.psu.edu/resources/PDFs/Economic%20Impact%20of%20Marcellus%20Shale%202009.pdf (accessed 1 April 2013).

Kilkenny, Maureen and Mark D. Partridge. (2009) "Export Sectors and Rural Development." *American Journal of Agricultural Economics*, Vol. 91, pp. 910–929.

Kleinhenz & Associates. (2011) "Ohio's Natural Gas and Crude Oil Exploration and Production Industry and the Emerging Utica Gas Formation."

Kraybill, David S. and Jeffrey Dorfman. (1992) "A Dynamic Intersectoral Model of Regional Economic Growth." *Journal of Regional Science*, Vol. 32, No. 1, pp. 1–17.

Michaels, G. (2010). "The Long Term Consequences of Resource-Based Specialisation." *The Economic Journal* 121, 31–57.

Oldham, Jennifer. (2012) "North Dakota Oil Boom Brings Blight with Growth as Costs Soar." Bloomberg.com. www.bloomberg.com/news/2012–01–25/north-dakota-oil-boom-brings-blight-with-growth-as-costs-soar.html (accessed 20 October 2013).

—— (Feb, 2012) "North Dakota's Oil Boom Strains Its Infrastructure." Bloomberg Businessweek. www.businessweek.com/magazine/north-dakotas-oil-boom-strains-its-infrastructure-02022012.html (accessed 20 October 2013).

Paleontological Research Institution (2012). "Understanding Drilling Technology." Marcellus Shale Issue 6. www.museumoftheearth.org/files/marcellus/Marcellus_issue6.pdf (accessed 20 October 2013).

Papyrakis, Elissaios and Reyer Gerlagh. (May, 2007) "Resource Abundance and Economic Growth in the U.S." *European Economic Review*, Vol. 51, No. 4, pp. 1011–1039.

Partridge, Mark D. and Dan S. Rickman. (2003) "The Waxing and Waning of U.S. Regional Economies: The Chicken-Egg of Jobs Versus People." *Journal of Urban Economics* 53: 76–97.

—— (2010) "CGE Modeling for Regional Economic Development Analysis." *Regional Studies*, 44: 1311–1328.

Randall, J. and G. Ironside (1996) "Communities on the Edge: An Economic Geography of Resource Dependent Communities in Canada." *Canadian Geographer*, Vol. 40, pp. 17–35.

Rodriguez, Francisco and Jeffrey Sachs. (1999). "Why Do Resource-Abundant Economies Grow More Slowly?" *Journal of Economic Growth*, Vol. 4, No. 3, pp. 277–303.

Rosenthal, Stuart S. and William C. Strange. (2001) "The Determinants of Agglomeration." *Journal of Urban Economics*, Vol. 50, pp. 191–229.

Sachs, Jeffrey D. and Andrew M. Warner. (1995) "Natural Resource Abundance and Economic Growth." NBER Working Paper No. 5398. National Bureau of Economic Research, Cambridge, MA.

Shactman, Brian A. (2012) "The Downside of N. Dakota's Oil Boom." CNBC. www.cnbc.com/id/46205381/The_Downside_of_N_Dakota_s_Oil_Boom (accessed 20 October 2013).

Simon, Curtis. (1998) "Human Capital and Metropolitan Employment Growth." *Journal of Urban Economics*, Vol. 43, No. 2, pp. 223–243.

Stabler, J.C. and M.R. Olfert. (2002) "Saskatchewan's Communities in the 21st Century: From Places to Regions." Regina, SK: Canadian Plains Research Center, University of Regina.

U.S. Bureau of Economic Analysis. Personal Income and Employment Summary. www.bea.gov/iTable/iTable.cfm?reqid=70&step=1&isuri=1&acrdn=5#reqid=70&step=1&isuri=1 (accessed 20 October 2013).

U.S. Bureau of Labor Statistics: Local Area Unemployment Statistics. www.bls.gov/lau/data.htm (accessed 20 October 2013).

—— Current Employment Statistics. www.bls.gov/ces/ (accessed 20 October 2013).

U.S. Energy Information Administration. (Dec, 2012) "AEO2013 Early Release Overview." www.eia.gov/forecasts/aeo/er/early_production.cfm (accessed 23 June 2014).

—— (Jun, 2012) "Emerging Oil and Gas Supplies: Future prospects for Oil and Gas Production." www.eia.gov/pressroom/presentations/staub_06272012.pdf (accessed 20 October 2013).

USC Global Energy Network. (2013) "Powering California: The Monterey Shale and California's Economic Future." http://gen.usc.edu/assets/001/84787.pdf (accessed 20 October 2013).

Van der Ploeg, F. (2011) "Natural Resources: Curse or Blessing?" *Journal of Economic Literature* 49(2): 366–420.

Van der Ploeg, F. and A.J. Venables. (2012) "Natural Resource Wealth: The Challenges of Managing a Windfall." *Annual Review of Economics*, 4: 315–337.

Vekshin, Alison and James Nash. (2013) "California Fracking may Boost State Economy 14%, USC Says." Bloomberg News. www.bloomberg.com/news/2013–03–14/california-fracking-may-boost-state-economy-14–usc-says.html (accessed 20 October 2013).

Weber, Jeremy. (2012) "The Effects of a Natural Gas Boom on Employment and Income in Colorado, Texas, and Wyoming." *Energy Economics*, Vol. 34, No. 5, pp. 1580–1588.

Weinstein, Amanda. (2014) "Labor Market Restructuring in the Shale Boom." Forthcoming in the *Journal of Regional Analysis & Policy*.

Weinstein, Amanda and Mark Partridge. (2011) "The Economic Value of Shale Natural Gas in Ohio." The Swank Program in Rural-Urban Policy Summary and Report.

3 Confronting an Uncertain Future

How U.S. Communities are Responding to Shale Gas and Oil Development

Susan Christopherson and Ned Rightor

Introduction

When people think about natural resource extraction, they think of places that are sparsely settled and far from cities and suburbs. The U.S. has a rich lore of "boomtowns" and "ghost towns," yet people rarely connect this history—and the boom–bust cycle it depicts—to contemporary resource development (Cortese and Jones 1977; Freudenberg and Wilson 2002; Kassover and McKeown 1981). Today's natural gas and oil development using horizontal drilling and high volume hydraulic fracturing (HVHF, commonly referred to as "hydrofracking" or just "fracking") is both similar to and different from our previous experience. It will produce the same cycle of boom and bust at the local level, but it is occurring nationally on a scale that raises policy concerns for at least twenty-eight states,[1] thousands of local governments, and for the federal government as well. Significantly, drilling is occurring or may occur in a wide variety of landscapes—in or near major cities, in residential neighborhoods, in semi-rural environments, and in isolated rural communities. Furthermore, the risks of shale development extend outward to communities from which drilling materiel and water are drawn, those located along the pipelines, roads and rails en route to and from the drill sites, and those through which wastewater is transported for disposal or the gas and oil flows to market. The controversial mining of 'frac sand' in the Midwest (Karnowski 2012), or the earthquakes caused by an injection well near Youngstown, Ohio (Choi 2013), or the explosion of a unit train carrying shale oil in Lac-Megantic, Quebec—all places far from the drilling fields—illustrate just how consequential these risks can be.

However, the same is true of the distribution of benefits, and Youngstown again provides an illustration: just two miles from that injection well, V&M Star is repurposing a mill shuttered thirty-four years ago by Youngstown Sheet and Tube to produce seamless pipe for the burgeoning hydrofracking industry infrastructure (Niquette and Varghese 2012). This vignette captures two realities of shale gas development—that job creation and economic benefits, as well as risks, occur all along the supply and distribution chain, not just within the drilling regions, and also that drilling regions may not see job creation and economic benefits proportionate to their risks. Risks are localized, and frequently long-lasting (e.g. public health effects, community social-psychological disruption) (Brasier et al. 2011; Jacquet 2013; Perry 2012). Benefits last only as long as new wells are being drilled and new investment is occurring, and benefits follow the money. When it comes to jobs and indirect or induced economic benefit, it is not where the activity takes place but where the money is spent that counts.

On a national scale, the discovery and exploitation of large shale gas and oil deposits in many areas of the country has been welcomed as a source of jobs in a period of long-term economic stagnation, and touted as a source of lower energy prices for consumers and energy security for the U.S.[2] The shale gas and oil boom is stimulating the construction of an energy infrastructure based on natural gas, despite increasing skepticism among industry leaders about its duration (Ayres 2013; Chazan 2013).

At a regional or local level, the calculation of costs and benefits is even more complicated. State and local policy makers are beginning to confront a list of risks, social and economic as well as environmental. Some may arise in the immediate "boom" phase; others pertain to the long term, after the boom dissipates. The extent of these risks differs from one region to another. What is consistent is the central dynamic of natural resource development: a cycle of dramatic growth and then decline, sometimes precipitous, over which policy makers have usually had little control. The uncertainties presented by shale gas and oil development, and the wide variety of U.S. regions it may affect, have stimulated unprecedented public education and participation. They have also raised new concerns about the post-hydrofracking future that require new policy approaches.

One indication of that uncertainty is the strong local community response to shale gas and oil development. The issues that dominate local public discussion include environmental risks, but also extend to perceived threats to existing industries and to a highly valued quality of life. In the next sections, we examine some of the social and economic risks associated with shale development. We then describe results from a study of local government responses to HVHF that illuminate how communities in the Marcellus Shale perceive the risks they face and whether those risks will be addressed. In the final section, we look at how policy makers can approach the social and economic risks connected with shale oil and gas development, even with the limited systematic knowledge we have about those risks and the underlying uncertainty that characterizes unconventional fossil fuel development.

Some Key Economic and Social Dynamics of Contemporary Shale Gas and Oil Development

Easy Come, Easy Go: The Big Picture

Like any nonrenewable resource development, shale development does bring an economic "boom" to extraction regions, at least during the period when drilling sites and support facilities are set up and drilling takes place. As drilling companies move into a community, population flows in. Local expenditures rise on everything from auto parts to pizza and beer. There is a modest increase in jobs outside the extraction industry itself (Freudenberg and Gramling 1998; Marchand 2011) in construction, transportation, retail, hotels and restaurants, entertainment and services. Landowners receive royalty payments and have extra money to spend. The tax base may expand, providing a windfall for a local government. Research on actual employment impacts in resource development regions indicates that job projections are typically over-stated (Weinstein and Partridge 2011; Weber 2012), but notwithstanding the exaggerated estimates and guesstimates that abound concerning this increased economic activity, any activity is very welcome in light of the "great recession".

However, while a natural resource extraction boom may bring job and population growth for a few years, it also increases public service costs and the cost of living for local residents, and "crowds out" other industries. Shale development brings an additional level of uncertainty: if a number of U.S. states are engaged in shale gas and oil extraction, drilling rigs may move at short notice from one region to another, causing a series of economic disruptions as drilling starts up, shuts down, and then starts up again (Best 2009).

Boomtowns frequently experience social problems brought about by the influx of a transient population that follows the oil and gas industry rigs from one place to another. After the boom ends and the drilling crews and their service providers depart, the region may have a smaller population and a poorer economy than before the extraction industry moved in (Feser and Sweeney 1999). If this boom–bust cycle is combined with environmental damage, the long-term costs to regions hosting shale gas and oil extraction may be considerable.

What is certain is that: 1) natural resource development—including unconventional fossil fuel development—is positive for some segments of the population (mineral rights owners, some businesses) and negative for others (renters, landowners without mineral rights, businesses in competing industries); and 2) when the commercially viable resources are depleted, drilling ceases—either temporarily or permanently—and there is an economic "bust". The businesses and personnel connected to resource extraction leave the community (Christopherson and Rightor 2012). Population and jobs depart the region (Feser and Sweeney 1999).

More difficult to assess is how great the gains actually prove to be, how long they last, how widely distributed the benefits are, how great the costs are—social costs as well as monetary ones—and whether or not those who benefit also pay the costs.

Lowering the Boom: Local Impacts

Documenting the extent and trend of social and economic impacts on local communities new to shale gas and oil development is difficult. With the exception of crime statistics, there are no data compilations that support the ready comparison of social and economic impacts across counties in conjunction with the progress of the drilling cycle. Data, if it is available at all, must be assembled county-by-county or agency-by-agency. What we do have are analyses of the social and economic impacts on counties in the Western States where shale gas and oil development has occurred for two decades, and anecdotal evidence from the northern tier counties of Pennsylvania where the most intense Marcellus Shale drilling has been under way since 2008. While not a precise forecast, this accumulating body of evidence at least provides a picture of what more rural or urban-adjacent localities should anticipate with shale gas and oil extraction.

Data definitively documenting how localities are affected by HVHF shale gas and oil development is currently unavailable because neither the states nor the federal government have been willing to collect it. The absence of data on impacts has hampered states' abilities to realistically assess the costs of HVHF development and impose impact fees or taxes to compensate for losses. However, despite the absence of statistical data, there is a large literature documenting similar social and economic impacts among localities in different shale plays. This literature includes environmental impact statements, public policy reports, academic journal articles and eyewitness accounts by journalists. In some cases there are multiple accounts of social and economic impacts in the same area, such as Sublette County, Wyoming or Williston, North Dakota, at different points in time. This literature forms the evidence used in this chapter.[3]

The consistent theme in this now extensive literature is that state and local governments—counties, cities, townships, villages—are subjected to a wide range of demands for new services or increased levels of service, and that the administrative capacity, staffing levels, equipment, and outside expertise needed to meet those demands are beyond what has been locally budgeted.

The drilling phase of shale gas development usually depends on an out-of-state workforce (Jacquet 2011). Although resident workers may be employed during the drilling phase as truck haulers or in service and construction jobs, even these jobs may be filled by workers who move into the drilling area while maintaining a permanent residence in another state. This in-migration of transient workers has been exacerbated by the great recession in the U.S.

and the paucity of job opportunities elsewhere in the nation. In the case of the drilling workforce itself, this means a sudden influx of young men—some with families, many without. Some will be experienced gas field veterans, while others will be drawn from other places to the boom and the prospect of work.

In Sublette County, Wyoming, for example, as the number of gas wells drilled *per year* climbed from one hundred in the year 2000 to more than five hundred in 2006, the population of Sublette County swelled by 24 percent. During that same period Wyoming's population overall grew by just 4 percent, indicating that workers and their families were flocking to the area to meet the new labor demand. The most dramatic increase in population came from teens and young adults aged fifteen to twenty-four, and even as adults aged twenty-five to forty-four were decreasing statewide, they were increasing in Sublette County. Indeed, all cohorts of working age adults increased more rapidly in Sublette County than statewide (Ecosystem Research Group 2008: 20–24).

According to Jacquet (2009), this short-term population influx creates significant demands on public services. Add enough young men with hard jobs, disconnected from family and community ties and seeking "a good time", and both legal and illegal forms of entertainment prosper. Therefore, communities needed more police. Traffic on major roads increased, as did the number of traffic accidents, the number of emergency room visits, and the demand for emergency response services. In addition, local schools experienced increased demand as some workers entering the region enrolled their children. Further, as the demand for all manner of goods and services increased and local businesses sought to exploit the boom, prices went up—not just for temporary residents, but for long-time local residents as well; Jacquet found that local prices in Sublette County increased by twice the national rate over a six-year "boom" period.

The price inflation characteristic of shale boom regions especially affects rental housing. Evidence from across contemporary shale plays indicates that rents rise dramatically in drilling areas. Local long-term renters who cannot afford their apartments are displaced, and may seek housing assistance from local government. Hotels and motels, vacation rentals, RV parks, and campgrounds fill up with transient gas drilling workers, and drilling companies may resort to the construction of storage-facility-like "man camps" to alleviate the housing shortage (Rumbach 2011; Rubinkam 2010).

Williston, North Dakota is an isolated prairie town where a shale oil and gas boom is now under way. The city has been inundated by people from all over the U.S. looking for work, and while in-migrants to Williston frequently find work, they have nowhere to live. The homelessness rate in the city rose above 19 percent in 2011, with many people living for long periods in temporary quarters. Unfortunately, Williston has experience with the boom–bust cycle of oil and gas development, and that experience has discouraged investment in the housing needed for this workforce. Local interviews indicate that: "Developers have been slow to build more apartments, largely because they got stung by the region's last oil boom that went bust in the 1980s" (McPherson, 2011).

And, in an example of "crowding out", the increased demand for accommodations may benefit hotel and motel owners and local restaurants, but it hurts other local businesses, as there are few rooms available for a more traditional clientele: business travelers, recreation seekers, and tourists.

While reliable regional data on the numbers of transient workers are difficult to come by, one Wyoming drilling community estimated that they would easily increase the town's census by 20 percent or more (Town of Marbleton 2007). In the long run, given the population declines suffered by many communities in potential drilling regions, an influx of new people might be welcome if the newcomers like the area and decide to stay. Indeed, the small state of Wyoming has seen population increases and an unemployment decline over the past decade, especially in communities near gas drilling areas. Perhaps unexpectedly, however, a rapid increase in drilling activity is not always associated with a commensurate increase in population *resident* in the counties where the drilling occurs. An analysis of population change in core natural gas drilling counties in the Marcellus Shale during the first decade of the 2000s indicates that the resident population in these largely rural counties has grown marginally, if at all (Christopherson and Rightor 2011).

There are various reasons that population growth does not occur in these core counties, but the most frequently cited are the absence of services, the higher cost of living, and the lower quality of life in an industrialized environment. A reporter interviewing drillers working in the northern tier of Pennsylvania about why they reside across the border in southern New York captured the reason in one quote:

> 'There is nothing there—there's no entertainment, there's nothing to do,' he said of that Pennsylvania locale as he sipped a margarita. 'Chemung County' (in New York), Mr. Cullen added, 'is where we spend our money.'
>
> (Navarro, 2012)[4]

Spreading the Cost: Regional Impacts

The impact in New York of gas development in rural Pennsylvania is but one example of how the economic and social impacts of shale gas and oil development—both positive and negative—are likely to be felt not only locally but regionally, affecting cities and counties in areas adjacent to the drilling localities and beyond.

Well sites are not the only feature of the industrialization brought about by shale gas development. Water extraction sites must be developed. Sand must be mined. These inputs and a variety of chemicals must be transported by trucks—many, many, heavy trucks—to the well pads. After extraction, gas flows from the well sites to the main transmission lines via a network of pipelines, compressor stations, and storage facilities. Oil moves by pipeline, or frequently by rail car, to the refineries. Flowback and produced water from the wells has to be transported to treatment facilities, which must be equipped to handle the increased volume and particular array of toxic and non-toxic wastes, or to injection wells. The facilities required will be located where geological or logistical factors dictate, but these operations may touch communities hundreds of miles from the drilling regions, often in another state.

Among the support facilities needed by a drilling region are man camps (caravan sites for transient workers), staging sites, water extraction sites, landfills for drill tailings, water treatment facilities, pipelines, compressor plants, gas storage facilities, and railroad spurs. These facilities create a wide range of potential environmental, economic, and social stressors, all of which have implications for the regional economy and existing industries, particularly tourism and agriculture (McGowan 2011; Rumbach 2011 Adams and Kelsey 2011).

For example, apart from the dangers inherent in a widespread network of pipes full of methane or in high-pressure equipment generally, noise is a major concern related to compressor stations: they produce noise levels in the 85 to 95 decibel range. These levels are at or above the U.S. Occupational Safety and Health Administration (OSHA) threshold of safety for an eight-hour day, and compressors work a twenty-four-hour day. Environmental stressors can have an effect on nearby citizens, adjacent property values, and on other industries in the vicinity, including those in nearby cities.

Another example of the impact shale development facilities may have on a rural region is provided by the proposed gas storage facility in the Finger Lakes region of New York State, a major area for tourism because of its scenic beauty, small towns and vineyards. This facility is being planned by Inergy Midstream, LLC for the former US Salt plant just north of Watkins Glen, New York, with underground storage for 1.45 billion cubic feet of natural gas. The new owners propose to add a liquid propane storage facility holding up to 88.2 million gallons, also underground, plus two large brine ponds on the surface.

The site for this major facility is near the intersection of two gas transmission pipelines, and as a salt mine it is an appropriate natural gas storage site. However, Watkins Glen, in largely rural Schuyler County, is not part of the 'fairway'—the purported 'sweet spot' for Marcellus drilling in New York—so it is not likely to obtain local tax revenue from well production. Whatever the plant may contribute in the way of local taxes, Watkins Glen currently depends on revenue from Finger Lakes tourism, attendance at its famous auto races, the local wine industry and agriculture.

Industrialization for All: Remote Impacts

The industrial inputs to the drill site may come from rural areas far from the oil and gas extraction sites. Of particular note here is the connection between the growing use of HVHF technology and the proliferation of sand mining in western Wisconsin and eastern Minnesota, transforming what were once small rural towns whose economies centered on agriculture and tourism, and dividing the populace over what the environmental and public health hazards of silica dust might mean for them and for those industries (Karnowski 2012; Deller and Schreiber 2012).

Transportation of the product also has distant impacts. The shale oil pumped from North Dakota's Bakken Shale play travels across rural America every day in thousands of rail cars, in train sets up to 120 cars long along routes that bisect major populated areas, to the Port of Albany in New York. These unit trains, carrying millions of gallons of oil, create environmental and safety risks along the rail route and at transshipment sites. In May 2013 the Department of Environmental Conservation of New York State approved the transshipment of 1.8 billion gallons of oil per year from rail cars to barges on the Hudson River.

We're All In This Together

Although the regulatory focus has been on well sites, a similar set of questions applies to these industrial facilities and logistics and transport services. Who—the federal government, the state, or localities—is to regulate them, and monitor and enforce standards? Do they have the staffing and resources that will be required? How will the funds to support those efforts be provided?

In the U.S., regulation of this extensive industrial infrastructure is likely to occur at a level of government above that of the locality, (through a state public utilities commission, for example), but the balance between federal, state, and local regulation is in flux as pivotal cases work their way through the courts in each state (Negro 2012). Strong "home rule" states like Texas permit localities a greater role (Fullenwider 2010), while the Commonwealth of Pennsylvania's Act 13 precludes all local regulation (Christopherson, Frickey and Rightor 2013). Court decisions in New York, West Virginia and Ohio (Ohio Ninth Appellate District Court of Appeals 2013, for example) have attempted to prescribe and proscribe what types of local ordinances and administrative requirements are permissible under state law.

Depending on the state, localities may have a role in the permitting of pipeline routes along city/county rights of way, or the regulation of road use. Local government may also require filings and notice to abutters, and demand incident reporting and filing of as-built drawings for emergency planning. For compressor stations, local regulation may be able to establish setbacks, maximum noise levels, fencing and landscaping requirements, and enhanced standards for units adjacent to residential areas. Flowback fluids from the hydrofracking process or the produced water from producing wells, if not reused, must be removed from the well sites by trucks and transported to treatment facilities or injection wells. While it is regulated by the federal EPA and/or state agencies, this traffic and these facilities may also be subject to permit or construction standards that are set or implemented at the local level.

All of these local or regional activities require expertise, administration, monitoring, and enforcement capacity, and all entail planning and public administration costs.

Taken together, the elements of this new industrial landscape spread far beyond the jurisdictions where drilling is taking place or production is being generated. However, severance taxes and other tax revenue schemes that are supposed to compensate communities for dealing with the attendant risks and impacts of shale gas and oil development are devised at the state level. Some yield revenue only for the state (e.g. California, Ohio). Some have discretionary revenue sharing that may route a portion of the proceeds to local governments (e.g. Colorado, Wyoming). Some permit local taxation as well (e.g. Texas). At least one (New York) supports local taxation only. Some (e.g. Pennsylvania) allocate arbitrary "impact fees" (Allegheny Conference on Community Development 2009).

None of these compensation schemes are based on an analysis of actual or potential costs to localities. Many have loopholes that favor the oil and gas producers, or lags in payment that leave localities in arrears (see Lepori in Christopherson 2011). Most are based on the volume or the value of the gas or oil produced, and they are frequently allocated according to the *place* where it was produced – i.e. the drilling region – which does not serve the needs of localities impacted by facilities or operations upstream or downstream of the actual extraction of the resource. If you have impacts but no drilling, you are often on your own.

* * *

These descriptions of some of the economic and social impacts on localities provide a sense of the types of risks communities may face over the shale gas and oil development cycle. They also demonstrate that the risks faced by communities are likely to differ from place to place. Variations in location, climatic conditions, local economies, and types of shale-development-related operations and facilities, as well as differences in local government capacity, make predictions of the consequences difficult. Furthermore, the difficulty of doing a systematic analysis of risks is sometimes exacerbated by state policies regarding data gathering and transparency.

However, these limitations do not eliminate the need for an effective risk governance framework that will enable localities to anticipate what is likely to occur, and help local and state officials design both preventive and ameliorative actions to reduce the impact of the social and economic consequences of shale development.

What Did Local Officials Tell Us About Community Perceptions of These Risks?

In 2012 we conducted a systematic study of local government responses to HVHF shale gas development in the Marcellus Shale states. We first developed a database of 298 localities or counties that had passed resolutions or legislation on shale gas development in the four states that contain most of the Marcellus and Utica shale plays: New York, Ohio, Pennsylvania, and West Virginia. Of the communities we identified, 266 were in New York (which has not yet authorized HVHF natural gas development) and Pennsylvania (which has). We then selected a stratified sample of communities in those two states and conducted fifty-three structured interviews with the highest-ranking public official or his or her designee in each community. These interviews obtained information on the process of decision-making, on the critical issues discussed in public meetings, and on community expectations regarding oil and gas industry practices and state regulation/monitoring of the industrial activities associated with HVHF. Although environmental issues topped the ranking of community concerns, particularly effects on water supply and water quality, localities that had taken legislative action were also concerned about public costs associated with increased traffic and road damage, and a variety of disruptions to local life.[5]

It is these costs—both monetary and to the quality of life—that are likely to be felt outside the immediate drilling areas: in communities from which drilling materiel and water are drawn, those on the roads and rails en route to the drill sites, or those through which wastewater is transported for disposal or through which the gas and oil flows to market.

The Deliberative Process

In the communities taking action to restrict HVHF that we surveyed in New York and Pennsylvania, the deliberative process was extensive. Of the forty-five communities responding to a question about the extent and content of official public meetings dedicated to discussing shale gas issues since 2008, twenty-two (49 percent) had between one and three public meetings or hearings, nine (20 percent) had between four and eight public meetings/hearings, and thirteen (29 percent) had more than eight. By contrast, in the New York communities we interviewed who passed resolutions in favor of state control (six communities), four had no public meetings. Although that sample is small, the minimal public involvement in these resolutions is substantiated by press accounts about a larger number of communities (Reilly 2012).

Community Concerns

Interviewees indicated that residents who testified during the meetings that preceded the legislative decision expressed concerns about a range of issues. Residents had concerns about environmental impacts from shale gas development, but they also raised public health issues, traffic congestion, and how the industrialization of the region would affect their quality of life. Not surprisingly, there was more discussion of potential community disruptions in communities that passed restrictive legislation, both in New York and Pennsylvania, whereas in communities that passed resolutions supporting shale gas development the focus was more on community benefits, although road maintenance and traffic congestion were mentioned as concerns.

Community meetings also included discussion of the benefits of shale gas development in places that passed restrictive legislation (but only in New York). Officials in the thirty-three "restrictive" communities we surveyed in New York were asked to rank the top three issues discussed in their deliberations. Out of a total of ninety-five entries, a benefit from shale gas drilling received a top-three ranking ten times. Among the benefits mentioned were tax benefits, economic benefits to local citizens, economic benefits to local businesses, and induced commercial development. The most frequently cited benefit was economic benefits to local citizens (mentioned six times). In Pennsylvania, our survey respondents did not list any of these benefits as being among the top three issues discussed in their public meetings.

We probed which concerns were most significant in driving action, as indicated by which issues were the most discussed during public deliberations. Our respondents indicated that the primary concern in thirty-seven out of the forty-five communities responding to the question among those that had passed restrictive legislation was with water issues. Water issues were among the top three issues for forty-four out of the forty-five. In two out of the six communities we interviewed that passed supportive legislation, water concerns were also mentioned as one of the top three items of discussion.

With respect to community impacts, the top concerns were road maintenance, public health, and traffic. Thirty-five of forty-nine respondents (71 percent) said road maintenance was one of their top three most-discussed community impacts during public deliberations. The second most commonly discussed community impact during public deliberations was public health, with twenty-seven of forty-nine respondents (55 percent) listing public health in their top three. Finally, traffic issues and traffic congestion were the third most discussed community impact, with twenty-five of forty-nine respondents (51 percent) listing this issue among their top three most discussed prior to legislative action.

Regulatory Preferences

Interviews with political leaders in our sample of New York and Pennsylvania communities that have enacted legislation or resolutions on HVHF indicate that the issue of shale gas development is important to local decision-makers, but equally important is the ability of the community to control its own destiny. When asked what level of government should control shale gas extraction, forty-three of fifty respondents (86 percent) indicated that local government should play a substantial role in regulating shale gas drilling. Twenty-four respondents indicated a preference for local control only, while nineteen favored a multi-level distribution of regulatory authority that includes localities. Indeed, only one of the six communities we talked to that had passed resolutions in *support* of pursuing shale gas development under state regulation advocated state regulation exclusively.

Public Costs and Local Capacity

We asked about local capacity to respond to the needs and expectations of the public and of gas companies for services during the drilling phase. On a scale of 1 to 5, with 1 being good, 3 being "Stressed but OK", and 5 being "Overwhelmed", thirty-nine of fifty respondents (78 percent) rated their capacity as somewhere between "Stressed but OK" and "Overwhelmed". The median response was 3.5.

Communities are unsure about how any costs related to shale gas development will be covered. When asked how they anticipated that any increased demand for local services would be paid for, twenty-four of fifty respondents (48 percent) said they were not sure. Only fourteen (28 percent) of the respondents anticipated state aid in the form of impact fees. No respondent anticipated that state taxes would be used to pay for their additional costs.

Trust Issues

Beyond their perception of specific environmental, economic or social risks lay a concern that those responsible for mitigating these risks or monitoring, assessing, and ameliorating any damage may be unwilling or unable to do so. We asked our interviewees four questions about their degree of confidence that the gas industry or their state government would protect the interests of affected communities, rated on a scale of 1 ("Very confident") to 5 ("Not at all confident").

- Asked about their confidence that the natural gas industry would protect the environment, health and safety of affected communities, thirty-one of forty-eight respondents (65 percent) said their confidence was low (4) or they were not at all confident (5).

- Asked about their confidence that the natural gas industry would protect the economic and social stability of affected communities, thirty-two of forty-nine respondents (65 percent) said their confidence was low (4) or they were not at all confident (5).
- Asked about their confidence that the state had the capacity to enforce environmental, health and safety regulations to protect affected communities, thirty-one of fifty respondents (62 percent) said their confidence was low (4) or they were not at all confident (5).
- Asked about their confidence that their state would regulate drilling activity effectively to protect the economic and social stability of affected communities, thirty-three of forty-nine respondents (67 percent) said their confidence was low (4) or they were not at all confident (5).

We conclude that a lack of trust in those responsible for creating these risks and dealing with the harms is as much responsible for local community responses as the fear of the risks themselves.

A Framework for Policymaking to Address These Risks (Notwithstanding Their Complexity, Uncertainty and Political Controversy)

Some U.S. states where unconventional fossil fuel development is occurring—such as Colorado, West Virginia and Texas—already have major fossil fuel extraction industries. Others—such as New York, Maryland, Virginia, and Illinois—have historical experience with fossil fuel extraction, but not with the types of extraction that transform regional economies, or that require widespread industrial facilities and extensive alteration of the landscape. In all the states and localities where HVHF drilling is a geological possibility, the social and economic risks associated with HVHF have been controversial, and the calculation of costs and benefits difficult to determine. People living in or near areas where extraction may occur are concerned with how an influx of transient workers will affect their cost of living and quality of life. They want to know about how hydraulic fracturing will affect existing industries. They want to know how the public costs associated with shale gas development, such as road repair or emergency services, will be paid. In all these cases we have sufficient information to indicate that risks are present, but insufficient information to predict their extent and cost.

Contemporary evidence from community and regional case studies is limited in its ability to tell state and local policy makers exactly what to expect from hydraulic fracturing as it moves through the development and extraction cycle. The variety of types of places affected, the differences in legacy effects from previous resource development cycles, differences in land ownership patterns, and differences in local and state government policy, necessarily mean that the extent of social and economic impacts will differ between extraction regions. What we can be confident of, however, is that natural resource extraction is characterized by a boom–bust cycle that is most acutely felt at the local level, and that the greater the scale and pace of drilling (the number of wells drilled, how quickly, and how consistently over time), the more intense will be the social and economic impacts on a region during the boom portion of the development cycle. Some communities are able to recover more rapidly than others because of their location in suburbs with other industrial employers or as gateways to national parks (Brown et al. 2005).

To secure the long-term economic sustainability of regions and communities affected by HVHF natural gas and oil development, state and local policy makers need to address the risks in several ways.

Analysis

First, given the differences among regions that could be affected by unconventional fossil fuel development, policy makers need to conduct a thorough analysis of how their state, region, or community may be affected. The full extent of potential social and economic risks needs to be acknowledged and addressed through regional and state policy. This includes communities anywhere along the industry's supply lines, waste disposal operations, or oil and gas transportation systems. Both historical and contemporary evidence indicates that unconventional fossil fuel development will entail an industrialization process in which social and economic risks are created far beyond the well site.

As for drilling regions, community dialogue and education about the risks can help public officials weigh the costs and benefits for a particular community or region. For example, regions with an important tourism industry may want to consider the risk of losing a substantial piece of that industry to "crowding out" or industrialization that damages their "brand". Drilling's competition for transportation services or a limited labor supply in the short term may hurt manufacturers and other employers of moderate skill workers that have committed to the area for the long term. Hardest to gauge is "opportunity cost": the proportion of new enterprises that will shy away from locating in a drilling region because of cost of living, labor supply, social disruption, or "image" concerns.

Control the Pace and Scale of Development

Second, government officials need to anticipate the natural resource development cycle, and recognize that effective governance requires controlling the pace and scale of development. Moderating the pace and scale of development can reduce adverse "boom"-related impacts in the short run, enable the community to adapt, and promote longer-term economic sustainability by controlling negative impacts on other local or regional industries.

Local zoning regulations, comprehensive planning requirements such as those currently proposed in Maryland,[6] and state permitting regulations can all be used as tools to limit the pace and scale of drilling while not preventing shale development altogether, and reduce the impact on local and state government services, public infrastructure, and other industries.

Cover the Costs

Third, to ensure that states cover the real costs of shale development and secure the long-term economic viability of affected regions, state and local policy makers need to jointly develop revenue sources and/or revenue sharing mechanisms that compensate communities for the uptick in need for facilities and services during the boom, and economic development programs that will capture and extend private investment to weather the decline (see Lepori in Christopherson 2011).

- *Create a tax that effectively pays for the short- and long-term costs of shale development.* States can impose a severance tax without risk of reducing production or industry jobs. If a state has a severance tax that is too low, shale extraction will produce a significant amount of additional government costs without commensurate fiscal benefits.
- *Distribute tax revenue predictably and fairly between state and local governments.* There are many ways to allocate revenue that are aligned with the costs of shale gas development. Regardless of the exact distribution, the primary purpose of a severance tax is to cover costs borne by the local and county governments affected.
- *Limit deductions and exemptions.* Many states have relatively high tax rates, but so many tax loopholes that the effective tax rate does not cover the cost of administering it, or the short- and long-term costs of shale development. Constructing a tax that is straightforward and simple makes compliance easier for gas producers and tax officials. Furthermore, because the structure of the tax determines how volatile it will be, exemptions and loopholes should be minimized.
- *Establish a permanent fund.* A permanent fund is the most effective way to promote long-term economic development. For example, every state in the intermountain west invests in a permanent fund. The permanent fund serves to protect the state against future recessions and revenue volatility, and enables ongoing fiscal benefits from the depletion of a non-renewable natural resource.

Plan Ahead

Fourth, local officials need to plan for both the boom and the bust that accompanies resource development. Well-documented baseline data gathered *before* the boom phase is a necessary prerequisite for local, county, or state governments to "price" the uptick in need for facilities and services, and the additional costs of social disruption and impacts on existing industries resulting from shale development.

Then, when new drilling falls off and as production declines—and tax receipts, royalty payments, business income, and jobs with it—affected regions may find themselves with overcapacity. This can be a period of steep decline in population and tax base, or simply one of significantly slower growth. Foresighted infrastructure planning and financing can help mitigate the stresses. Flexible fiscal tools can enable localities or the state to accommodate fluctuating revenues and service demands. Budgeting to build reserves and support economic development can help communities weather the period after extraction ends.

Engage and Cooperate

Finally, good policymaking to protect communities from the social and economic risks of boom–bust development and to insure economic and environmental sustainability depends on good inter-governmental cooperation and citizen engagement. The goal should be to ensure that citizens trust the fairness of the policymaking process, and are justifiably confident that the costs and benefits of resource development will be appropriately and equitably distributed.

Notes

1 The latest U.S. Energy Information Administration map (www.eia.gov/oil_gas/rpd/shale_gas.pdf) shows "current" or "prospective" shale plays in twenty-seven states, but does not account for the Triassic Basin shales in North Carolina (http://geology.com/articles/north-carolina-natural-gas/). According to Davis (2011) and Wiseman (2013), twenty-seven states *currently* have identifiable deep shale or coalbed methane deposits *and* enough gas-producing activities to warrant regulation. Davis (2013) cites thirty-three gas-producing states overall, while America's Natural Gas Alliance (ANGA) "representing North America's largest independent natural gas exploration and production companies" identifies thirty-two (http://anga.us/why-natural-gas/abundant/shale-plays – .UkgdG4WkLbk). Again, however, this does not include North Carolina. Whatever the number of states that eventually experience HVHF shale *drilling*, the extraction of drilling materiel, the disposal of waste products, and the transportation of the gas and oil produced will affect many non-producing states.

2 Neither lower U.S. prices nor U.S. "energy independence" are assured. Absent a national energy policy that retains shale gas and oil produced in the U.S. for domestic use, the industry is free to export them to the highest bidder, and the price and availability of those products will continue to be determined by supply and demand on the world market. Further, to the extent that the gas and oil are used for electric power generation, the energy prices actually paid by U.S. consumers are determined by large electric utility companies. Customers in the states that have de-regulated their energy markets (including New York, California and Texas) pay higher rates for electricity, notwithstanding the price

or availability of the source of energy used to generate it (American Public Power Association 2013; Texas Coalition for Affordable Power 2012).

3 In addition to the references cited on particular topics, key sources include: Fuller 2007; Headwaters Economics 2009, 2012; Kelsey et al. 2012; King 2012; Putz et al. 2011; Perez-Burgos and Donaghy 2012; Pinedale Anticline Working Group 2006; Wilber 2012.

4 For a close-up of this border region between the "Northern Tier" counties of Pennsylvania and the "Southern Tier" counties of New York, see the map generated by the Susquehanna River Basin Commission at www.srbc.net/atlas/downloads/BasinwideAtlas/PDF/1404b_ABR.pdf. For an overview of the entire Marcellus Shale Formation area, featuring the state and county borders, see the map drawn by Timothy Murtha (using data from the National Atlas of the United States and the United States Geological Survey) at http://march.rutgers.edu/2011/11/marcellus-shale-the-cultural-landscape-of-natural-gas-in-pennsylvania/.

5 A full report on our research results is available at www.greenchoices.cornell.edu.

6 The June 25, 2013 *Recommended Practices for Marcellus Shale Drilling Released for Public Comment* is available on the Maryland Department of the Environment website at: www.mde.state.md.us/programs/Land/mining/marcellus/Pages/MSReportPartII_Draft_for_Public_Comment.aspx (accessed 10 July 2013).

References

Adams, R. and T. Kelsey. 2011. "Pennsylvania Dairy Farms and Marcellus Shale, 2007–2010." Pennsylvania State University Extension, Marcellus Fact Sheet. Available at: http://pubs.cas.psu.edu/FreePubs/PDFs/ee0020.pdf (accessed 15 October 2013).

Allegheny Conference on Community Development. 2009. "Benchmarking Pennsylvania: A Summary of Severance Taxes on the Natural Gas Industry." Draft report, no author(s) listed, February.

American Public Power Association. 2013. *Retail Electric Rates in Regulated and Deregulated States: Update 2012*. Available at: www.publicpower.org/files/PDFs/RKW%5FFinal%5F%2D%5F2012%5Fupdate.pdf (accessed 17 October 2013).

Ayres, Robert U. 2013. "Shale Oil and Gas: The Contrarian View." *Forbes*, May 8. Available at: www.forbes.com/sites/insead/2013/05/08/shale-oil-and-gas-the-contrarian-view/ (accessed 15 October 2013).

Best, A. 2009. "Bad Gas or Natural Gas, The Compromises Involved in Energy Extraction." *Planning Magazine*, October: 30-34. Chicago IL: American Planning Association. Available at www.planning.org/planning/2009/oct/ (accessed 30 December 2013).

Brasier, K.J., M.R. Filteau, D.K. McLaughlin, J. Jacquet, R.C. Stedman, T. Kelsey, and S.J. Goetz. 2011. "Residents' Perceptions of Community and Environmental Impacts from Development of Natural Gas in the Marcellus shale: A Comparison of Pennsylvania and New York Cases." *Journal of Rural Social Sciences* 26(1): 32–61.

Brown, R.B., S.F. Dorius, and R.S. Krannich. 2005. "The Boom-Bust Recovery Cycle: Dynamics of Change in Community Satisfaction and Social Integration in Delta, Utah." *Rural Sociology* 70(1): 28–49.

Chazan, Guy. 2013. "Peter Voser says he Regrets Shell's Huge Bet on US Shale." *The Financial Times*, October 6. Available at: www.ft.com/intl/cms/s/0/e964a8a6-2c38-11e3-8b20-00144feab7de.html - axzz2hp70z6yL (accessed 15 October 2013).

Choi, Charles Q. 2013. "Fracking Practices to Blame for Ohio Earthquakes." Livescience.com. Posted September 4. Available at: www.livescience.com/39406-fracking-wasterwater-injection-caused-ohio-earthquakes.html (accessed 7 September 2013).

Christopherson, S. (et al.) 2011. "The Economic Consequences of Marcellus Shale Gas Extraction: Key Issues." *CaRDI Reports* 14, September. Ithaca NY: Cornell University Community & Regional Development Institute. Available at: http://greenchoices.cornell.edu/downloads/development/shale/marcellus/Economic_Consequences.pdf (accessed 29 March 2014).

Christopherson, S., C. Frickey, and N. Rightor. 2013. *A Vote of No Confidence: Why Local Governments Take Action in Response to Shale Gas Development.* Ithaca NY: Cornell University, Department of City and Regional Planning. Report in the CRP Working Paper Series. Available at: www.greenchoices.cornell.edu/downloads/development/shale/marcellus/Vote_of_No_Confidence_WP.pdf (accessed 29 March 2014).

Christopherson, S. and N. Rightor. 2011. "How Should We Think About the Economic Consequences of Shale Gas Drilling." Ithaca NY: Cornell University, Department of City and Regional Planning. Working Paper Series for *A Comprehensive Economic Impact Analysis of Natural Gas Extraction in the Marcellus Shale.* Available at: www.greenchoices.cornell.edu/downloads/development/shale/marcellus/Thinking_about_Economic_Consequences.pdf (accessed 29 March 2014).

—— 2012. "How Shale Gas Extraction Affects Drilling Localities: Lessons for Regional and City Policy Makers." *Journal of Town & City Management* 2(4). London: Henry Stewart Publications. Available at: www.greenchoices.cornell.edu/downloads/development/shale/Economic_Effects_on_Drilling_Localities.pdf (accessed 29 March 2014).

Cortese, C.F., and B. Jones. 1977. "The Sociological Analysis of Boomtowns." *Western Sociological Review* 8(1): 75–90.

Davis, Charles. 2011. "The Politics of 'Fracking': Regulating Natural Gas Drilling Practices in Colorado and Texas." Western Political Science Association 2011 Annual Meeting Paper. Available at SSRN: http://ssrn.com/abstract=1766675 (accessed 30 December 2013).

—— 2013. "Substate Federalism and Fracking Policies: Does State Regulatory Authority Trump Local Land Use Autonomy?" Presentation to the National Research Council workshop on Governance of Risks of Shale Gas Development, Washington DC, August 15–16. Available at: http://sites.nationalacademies.org/xpedio/groups/dbassesite/documents/webpage/dbasse_084412.pdf (accessed 30 December 2013).

Deller, Steven C. and A. Schreiber. 2012. *Frac Sand Mining and Community Economic Development.* Madison WI: University of Wisconsin, Department of Agricultural and Applied Economics. Staff Paper Series, Number 565, May.

Ecosystem Research Group. 2008. *Sublette County Socioeconomic Impact Study, Phase I Final Report, prepared for the Sublette County Commissioners.* Missoula MT: Ecosystem Research Group, January.

Feser, E. and S. Sweeney. 1999. *Out-migration, Population Decline, and Regional Economic Distress.* Washington, DC: US Department of Commerce. Report prepared for the US Economic Development Administration.

Freudenburg, W.R. and R. Gramling. 1998. "Economic Linkages in an Extractive Economy." *Society and Natural Resources* 11: 569–586.

Freudenburg, W.R. and L.J. Wilson, 2002. "Mining the Data: Analyzing the Economic Implications of Mining for Nonmetropolitan Regions." *Sociological Inquiry* 72: 549–557.

Fullenwider, Sarah (Senior Assistant City Attorney, City of Fort Worth, TX). 2010. "If We Knew Then What We Know Now: A Decade of Lessons Learned from Urban Drilling in Fort Worth." Webinar presented May 20. Pennsylvania State University Cooperative Extension, College of Agricultural Sciences, Marcellus Shale Educational Webinar Series, April–September 2010. Available at: https://meeting.psu.edu/p67821187/?launcher=false&fcsContent=true&pbMode=normal (accessed 30 December 2013).

Fuller, A. 2007. "Letter from Wyoming: Boomtown Blues." *The New Yorker*, February 5: 38. Available at: http://archives.newyorker.com/?i=2007-02-05#folio=002 (accessed 13 October 2013).

Headwaters Economics. 2009. *Fossil Fuel Extraction as a County Economic Development Strategy. Are Energy-focusing Counties Benefiting?* Bozeman, MT: Headwaters Economics. Available at: http://headwaterseconomics.org/pubs/energy/HeadwatersEconomics_EnergyFocusing.pdf (accessed 30 December 2013).

—— 2012. *Oil and Natural Gas Fiscal Best Practices: Lessons for State and Local Governments.* Bozeman, MT: Headwaters Economics. Available at: http://headwaterseconomics.org/wphw/wp-content/uploads/Energy_Fiscal_Best_Practices.pdf (accessed 30 December 2013).

Jacquet, Jeffrey B. 2009. *Energy Boomtowns & Natural Gas: Implications for Marcellus Shale Local Governments & Rural Communities.* NERCRD Rural Development Paper No. 43. University Park, PA: The Pennsylvania State University. Prepared for the Northeast Regional Center for Rural Development.

—— 2011. "Workforce Development Challenges in the Natural Gas Industry." Ithaca NY: Cornell University, Department of City and Regional Planning. Working Paper Series for *A Comprehensive Economic Impact Analysis of Natural Gas Extraction in the Marcellus Shale.* Available at: www.greenchoices.cornell.edu/development/marcellus/reports.cfm (accessed 29 March 2014).

—— 2013. "Risks to Communities from Shale Gas Development." Presentation to the National Research Council workshop on Risks from Shale Gas Development, Washington DC, May 30–31. Available at: http://sites.nationalacademies.org/xpedio/groups/dbassesite/documents/webpage/dbasse_083401.pdf (accessed 15 October 2013).

Karnowski, Steve. 2012. "Natural gas, Oil Boom Spurs Sand Mining in Midwest." Associated Press, January 6.

Kassover, J. and R.L. McKeown. 1981. "Resource Development, Rural Communities and Rapid Growth: Managing Social Change in the Modern Boomtown." *Minerals and the Environment* 3(1): 47–57.

Kelsey, T.W., A. Metcalf, and R. Salcedo. 2012. *Marcellus Shale: Land Ownership, Local Voice, and the Distribution of Lease and Royalty Dollars.* The University Park PA: The Pennsylvania State University. CECD Research Paper Series.

Available at: http://aese.psu.edu/research/centers/cecd/publications (aceessed 30 December 2013).

King, Pamela. 2012. "Booming Bakken Town Struggles to Adapt as 'Gillette Syndrome' sets in." *EnergyWire* website, June 4. Washington DC: Environment and Energy Publishing, LLC. Available at: http://www.eenews.net/stories/1059965247 (accessed 15 October 2013).

Marchand, J. 2011. "Local Labor Market Impacts of Energy Boom Bust Boom in Western Canada." Available from the author at the Department of Economics, University of Alberta, 7-29 HM Tory, Edmonton, AB, Canada, T6G 2H4.

McChesney, John and Ann Chambers Noble interviews by Neal Conan, with listener comments. 2013. "Gas, Oil Booms Bring Complications To Small Towns." *Talk of the Nation*. National Public Radio, February 11. Transcript available online at: www.npr.org/2013/02/11/171718769/gas-oil-booms-bring-complications-to-small-towns (accessed 15 October 2013).

McGowan, E. 2011. "Gas Drilling Turning Quiet Tourist Destination into Industrial Town." One of a seven-part series by the author for InsideClimate News, May 20. Available at: http://insideclimatenews.org/news/20110517/fracking-pennsylvania-natural-gas-drilling-marcellus-shale (accessed 10 October 2013).

McPherson, J. 2011. "Oil Boom Raises Rents in ND, Pushes Seniors Out." Associated Press, November 14. Available at: www.nbcnews.com/id/45292393/ns/us_news-life/t/oil-boom-raises-rents-nd-pushes-seniors-out/#.UffoxOBh5UQ (accessed 26 July 2013).

Navarro, M. 2011. "With Gas Drilling Next Door, County in New York Gets an Economic Lift." *New York Times*, December 27. Available at: www.nytimes.com/2011/12/28/nyregion/hydrofracking-gives-chemung-county-ny-economic-boost.html?pagewanted=all&_r=1& (accessed 24 December 2013).

Negro, S. 2012. "Fracking Wars: Federal, State and Local Conflicts over the Regulation of Natural Gas Activities." *Zoning and Planning Law Report* 35(2). Available at: www.rc.com/documents/negro_frackingwars_2012.pdf (accessed 18 October 2013).

Niquette, Mark and Romy Varghese. 2012. "Youngstown Opens Mills Again as States Jockey for Fracking Jobs." Bloomberg News Service, January 10. Available at: www.bloomberg.com/news/2012-01-10/youngstown-opens-mills-again-as-states-jockey-for-fracking-jobs.html (accessed 30 September 2013).

Ohio Ninth Appellate District Court of Appeals. 2013. Opinion in *State of Ohio ex rel. Jack Morrison Jr., Law Director, City of Monroe Falls, Ohio, et al. vs. Beck Energy Corporation, et al.* (State ex rel. Morrison v. Beck Energy Corp., 2013-Ohio-356). Available at: www.slideshare.net/MarcellusDN/oh-court-decision-striking-down-municipal-home-rule-for-oil-gas-drilling (accessed 18 October 2013).

Perez-Burgos, J. and K. Donaghy. 2012. "Regional Economic Impact Analyses of Natural Gas Exploration and Extraction in the Marcellus Shale." Ithaca NY: Cornell University, Department of City and Regional Planning. Working Paper. Available from the authors.

Perry, S.L. 2012. "Development, Land Use and Collective Trauma: The Marcellus Shale Gas Boom in Rural Pennsylvania." *Culture, Agriculture, Food and Environment* 34(1): 81–92.

Pinedale Anticline Working Group. 2006. Draft *Monitoring Plan and Report* by the SocioEconomic Task Group, March. Data contained therein is available at: www.sublette-se.org (accessed 15 December 2013).

Putz, A., A. Finken and G. Goreham. 2011. "Sustainability In Natural Resource-Dependent Regions That Experienced Boom-Bust-Recovery Cycles: Lessons Learned From A Review Of The Literature." Fargo ND: North Dakota State University. Report prepared for the NDSU Extension Service Center for Community Vitality. Available at: www.yumpu.com/en/document/view/4384777/sustainability-in-natural-resource-dependent-regions-boom-bust (accessed 18 October 2013).

Reilly, S. 2012. Pressconnects.com, July 10. Available at: www.pressconnects.com/article/20120710/NEWS01/207100393/Broome-towns-?nclick_check=1 (accessed 20 January 2013).

Rubinkam, Michael. 2010. "Chesapeake Unveils 'Man Camp' in Bradford County." *The Star Gazette*, November 18.

Rumbach, A. 2011. "Natural Gas Drilling in the Marcellus Shale: Potential Impacts on the Tourism Economy of the Southern Tier." Ithaca NY: Cornell University, Department of City and Regional Planning. Working Paper Series for *A Comprehensive Economic Impact Analysis of Natural Gas Extraction in the Marcellus Shale*. Available at: http://greenchoices.cornell.edu/development/marcellus/policy.cfm (accessed 29 March 2014).

Texas Coalition for Affordable Power. 2012. *Deregulated Electricity: A History of Retail Competition*. Available at: http://historyofderegulation.tcaptx.com/SB7-Report-2012.pdf (accessed 29 March 2014).

Town of Marbleton. 2007. *RE: Impact Analysis 07*. Letter to Sublette County Board of Commissioners (Marbleton WY).

Weber, J. 2012. "The Effects of a Natural Gas Boom on Employment and Income in Colorado, Texas, and Wyoming." *Energy Economics* 34(5): 1580--1588.

Weinstein, A. and M. Partridge. 2011. *The Economic Value of Shale Natural Gas in Ohio*. The Swank Program in Rural-Urban Policy Summary and Report, December. Columbus OH: The Ohio State University, Department of Agricultural, Environmental, and Development Economics. Available at: http://aede.osu.edu/research/c-william-swank-program/policy-briefs (accessed 30 December 2013).

Weisz, R. 1979. "Stress and Mental Health in a Boom Town." In J. Davenport & J. Davenport, II (Eds), *Boom Towns and Human Services*. Laramie WY: University of Wyoming, Department of Social Work.

Wilber, T. 2012. *Under the Surface: Fracking, Fortunes, and the Fate of the Marcellus Shale*. Ithaca NY: Cornell University Press.

Wiseman, Hannah J. 2013. "Evaluating and Enhancing the Capacity of the States to Govern Unconventional Oil and Gas Development Risks." Presentation to the National Research Council workshop on Governance of Risks of Shale Gas Development, Washington DC, August 15–16. Available at: http://sites.nationalacademies.org/xpedio/groups/dbassesite/documents/webpage/dbasse_084382.pdf (accessed 15 October 2013).

4 Climate Change, Energy and Agriculture

Interrelationships and Possible Policy Approaches

Bruce A. McCarl and Marta Wlodarz

Today one can hardly read a newspaper without a mention of climate change. While evidence of climate change is widespread, a uniform policy approach has not emerged. This chapter presents a research-based review of the issue, and discusses possible policy directions with an emphasis on energy, agriculture and forestry.

What is Happening

The world is facing increasing global temperatures (see Figure 4.1). Eleven of the twelve warmest years in Earth's recorded history have happened since 2000. The United States was at its warmest in 2012. Global sea level is rising, rainfall patterns are changing, and drought is becoming more common. Weather is becoming more extreme, leading to ever more costly events. For example, the drought in the U.S. Southwest in 2011 was estimated to cause a net loss of $7.6 billion (Texas AgriLife Extension Service 2012). We are in a period of rapid climatic change.

Causal Factors and Possible Mitigation

Recent findings suggest that carbon dioxide and other greenhouse gases (GHGs) are a driving force behind climate change. The theory is that GHGs trap solar radiation, thereby increasing the retention of heat in Earth's atmospheric system. Emissions are rising at an accelerating rate (Figure 4.2). Climate stabilization requires that emissions peak and then drop to low levels (IPCC 2007c). Through the United Nations Framework Convention on Climate Change (UNFCCC) society has a stated goal to reduce GHG emissions, but progress is slow. Unilateral action by any one country would not be effective; stabilization has to be a global endeavor (Fri et al. 2010).

Energy and agriculture are intimately involved. Globally, 69 percent of greenhouse gases originate from energy consumption (IPCC 2007a). About 14 percent (note that this includes energy emissions counted in the 69 percent total) arise from food production, and 17 percent from deforestation.

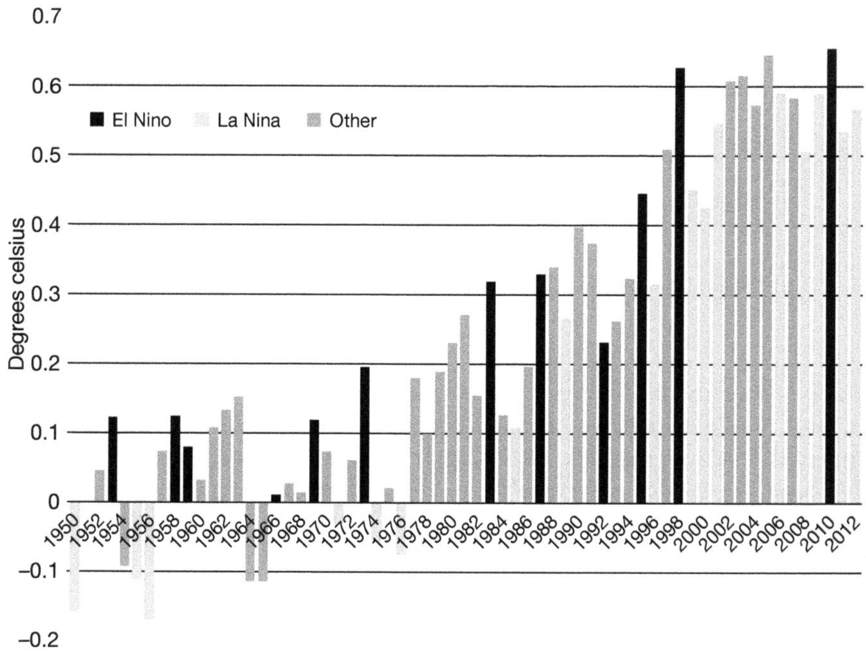

Figure 4.1 Temperature Evolution from 1950–2012.
Source: National Oceanic and Atmospheric Administration (NOAA), State of the climate, Global Analysis – Annual 2012 www.ncdc.noaa.gov/sotc/global/2012/13.

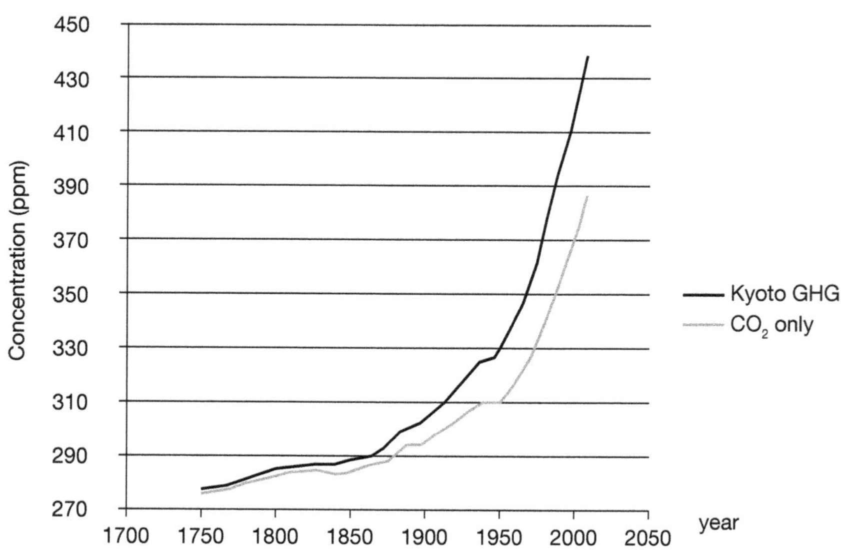

Figure 4.2 Expansion of GHG Emissions for Carbon Dioxide and all GHGs, 1750–2007.
Source: authors.

What is Projected

Given continuing emissions climate change is projected to continue, with further temperature and precipitation fluctuations accompanied by many other environmental changes. Figure 4.3 portrays temperature projections, with the top line depicting the current global emissions path and the other lines representing alternative emissions reduction scenarios. A continuation of substantial emissions would result in widespread disruptions including changes in fresh water supplies, reductions in snow pack, alterations in the location of agricultural crop production zones, altered frequency of extreme event incidence, increases in sea level, increased inundation of coastal infrastructure, altered incidence of pests and diseases, and changed ranges of birds and plants.

Effects of Climate Change

Climate change is not expected to have uniform implications. Instead, it will have differential effects in different regions of the world, as well as different effects for ecosystems, animals, plants and human health. We briefly discuss some of these effects below, but Intergovernmental Panel on Climate Change (IPCC) documents (IPCC 2007a, b, c and predecessors) provide much more evidence.

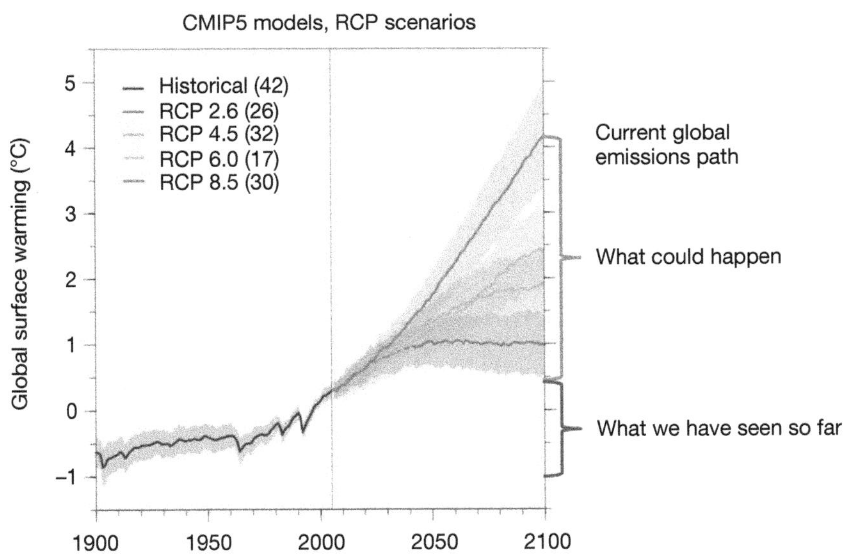

Figure 4.3 Global Temperature Changes from New Coupled Model Intercomparison Project Phase 5 (CMIP5) Climate Model Projections.
Source: Reto Knutti and Jan Sedláček, Nature Climate Change (2012).

Effects on Crops and Livestock

Effects of climate change on crop yields and livestock have been examined both in the United States and worldwide. Adams et al. (1990, 1995, 1999), IPCC (2007b) and Godfray et al. (2010) show differential agricultural effects across U.S. and other world regions, as do a number of other studies including the U.S. 2001 National Assessment (Reilly et al. 2001, 2002; IPCC 2007; Beach et al. 2010; McCarl 2011; Attavanich et al. 2013). Yield effects differ in part because of proximity to the poles. Butt et al. (2005) show yield losses in Mali, while the U.S. National Climate Assessment shows reductions in some southern areas for some crops (Reilly et al. 2001, 2002). Almaraz et al. (2009) and Olesen and Bindi (2002) concluded that climate change will benefit agricultural production in Canada and northern parts of Europe. The IPCC (2007b) provides much more evidence, generally showing reductions in lower latitudes and increases in upper latitudes. The same will undoubtedly be shown in the next assessment report. Unfortunately, subtropical areas (the location of many developing countries) have been identified as the most vulnerable (IPCC 2007b).

CO_2 will also affect yields (Long et al. 2006; Attavanich and McCarl 2011). Soybeans, rice and wheat (which are called C3 crops) benefit from increased atmospheric CO_2 while C4 crops (maize, millet, sugarcane and sorghum) are generally unaffected, although they can benefit under moisture stress.

In terms of livestock, productivity is affected through both direct and indirect means. Hotter temperatures alter the animal heat balance and consequently alter mortality, feed conversion, rate of gain, milk yield, and fertility (Hahn 1995, 2000; Mader et al. 2009). Appetite may also be affected (Adams et al. 1999). Heat waves have been found to increase death loss (Hahn and Mader 1997; Hahn et al. 2001). Altered water supply, land use, feed production, grass growth, incidence of parasites and pathogens and altered location of disease vectors are also factors. For example, Mu (2012) finds a correlation between climate change and the incidence of avian influenza.

Impacts on Extremes and Food Security

The IPCC Special Report on Extreme Events (2012) indicates a strong correlation between climate change and droughts, floods, heat waves, and extreme cold spells. According to the National Oceanic and Atmospheric Administration (NOAA), events like the 2011 drought in the southwestern U.S. are expected to be about twenty times more likely during La Niña years than they were in the 1960s due to realized climate change (NOAA 2013). Moreover, droughts are projected to be likely to intensify in many regions, especially in North and South America, Africa and Central Europe. As a result, food supplies might be severely diminished, especially in western and sub-Saharan Africa. Under unabated climate change, Downing (1991) indicates that droughts may cause Kenyan households to have insufficient food in 73 percent of the months. Ifejika Speranza et al. (2008) provide evidence that most famines in Kenya have been caused by droughts. Butt et al. (2005) estimate that climate change would increase the population at risk of hunger from 34 percent to between 64 percent and 72 percent. Lobell et al. (2008) find that South Asia and southern Africa are highly vulnerable to food insecurity caused by climate change.

Additionally, lower food supplies mean higher goods prices and increased difficulties for the poor. Qureshi et al. (2013) finds that changes in precipitation patterns will affect crop yields in Australia, and in turn will have an impact on food prices in the wheat, meat and dairy markets.

Vector-borne Diseases and Pest Control

Higher air temperatures, shorter freeze periods, more frequent heat waves and higher humidity can cause some insects, pests and diseases to be more prevalent (Rosenzweig et al. 2000; Diffenbaugh et al. 2008) with expansion into previously unaffected areas.

It is not only staple crops that will experience increased pest pressure. Schmitz and Shapiro (2012) find that the cacao tree is under threat from pests and fungal infections aggravated by climate change and the farmers' lack of access to pesticides and fertilizers. Powell and Logan (2005) suggest that climate change has been influencing forest insects, finding an intensification of the mountain pine beetle in North America plus a northward and higher altitude shift in population distribution. Changing climate also stimulates in-migration of invasive species or pests (Runyon et al. 2012; Sortel et al. 2013).

Larger pest populations mean more pesticide costs for the agriculture sector. Chen and McCarl (2001a) examined the relationship between climate change and pesticide costs. They concluded that increases in precipitation raise average per-acre pesticide usage costs for corn, cotton, potatoes, soybeans and wheat, while higher temperatures increase pesticide costs for corn, cotton, potatoes and soybeans but decrease the cost for wheat. Shakhramanyan et al. (2013) also found higher pesticide application levels and accompanying damage to human health and other externalities.

Sea Level, Coastal Areas and Agriculture

Climate change and sea-level rise (SLR) are inevitably interconnected (Robinson et al. 2012; Hanna et al. 2013). The IPCC (2007a) projected that global sea levels will rise by 0.23m to 0.47m by 2099 (under the A1B scenario[1]), purely from thermal expansion, Hanna et al. (2013) found a rise in the neighborhood of 3mm per year over the past few decades. Other larger sea-level increases have been suggested (Dasgupta et al. 2009). Rising sea levels means inundations of coastal areas and small islands. Agricultural production in low-lying areas, including rice production, will also be affected (Verbeek 2003; Rekacewicz 2006; Chen et al. 2012).

Water

Climate change can alter water availability, demand and quality. Rising temperatures decrease snowfall and snow storage in the mountains, altering runoff timing (Adams and Peck 2008). Hotter conditions increase plant respiration and evapotranspiration, raising irrigation water requirements. Additionally, water use is stimulated by higher levels of exposure to CO_2 (Adams et al. 1999). Water supply is altered by drought incidence, changes in precipitation amount and variability (Gleick et al. 2000). Some regions are expected to incur substantial reductions (IPCC 2007a; Seager et al. 2007, 2013). Observations show increased summer drought prevalence in the last thirty years in certain world regions—for example, in the southwestern U.S., Australia, and sub-Saharan Africa (USCCSP 2008; IPCC 2012).

An increase in nonagricultural water demand will occur under hotter temperatures, and lowered precipitation will increase water competition. For example, Chen et al. (2001b) and Cai and McCarl (2009) found that climate change increased demand for water resources from municipal and industrial users in Texas. Aquifer depletion will likely be greater in such circumstances (IPCC 2008; Wang et al. 2013).

Water quality can also be affected. Higher water temperatures and longer periods of low flows are projected to exacerbate water pollution from sediments, chemical contamination, dissolved organic carbon, pathogens and pesticides (IPCC 2008). In coastal areas, rising sea levels may have negative effects on storm-water drainage, sewage disposal and the intrusion of saline water into groundwater.

Policy Landscape

There are three broad policy approaches to climate change (see Figure 4.4). One involves investing in learning more about the issue. There are also two preventative approaches. These are commonly called adaptation and mitigation (IPCC 2007b, c). Mitigation involves actions aimed at reducing climate change greenhouse gas emissions, which are designed to limit future climate change. Adaptation involves reducing the negative effects of climate change and exploiting the positive ones.

Note that these three basic policy directions are not mutually exclusive and will all likely be used. Next we will discuss these policy directions at greater length, with a bias toward energy and agriculture.

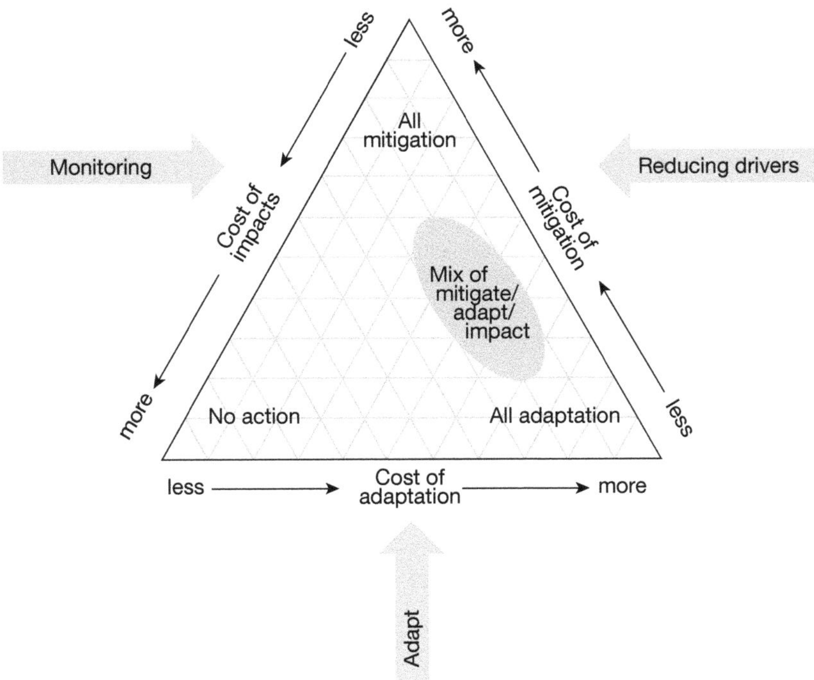

Figure 4.4 Three Elements of Total Burden of Climate Change.
Source: Parry et al. (2009).

Information Gathering

Society can fund activities that gather information on the degree of climate change and its observed consequences. This may involve large or small investments. It also may involve waiting for information that clarifies effects and causal mechanisms. This activity is essential and will reduce uncertainty regarding the need for mitigation and adaptation. More time dedicated to the observation of climate change impacts generally will help in making better decisions about funding allocation. For example, the U.S. Climate Change Science Program has the objective of conducting research and gathering necessary data to support policy decision making.

However, overreliance on such an approach can also lead to delayed actions, which may in turn cause substantial damage and some irreversibilities (Houghton et al. 1996). For example, the Global Coral Reef Monitoring Network estimated that taking no action against changing climate will cause \$350 million to \$870 million of annual coral reef-based loss in the Caribbean by 2015 (Corcoran et al. 2010). The coral reef loss has been translated into economic impact terms by estimating the values of selected goods and services derived from coral reefs in the Caribbean (for the year 2000) and estimating potential losses due to coral reef degradation (by 2015). Key sectors considered included fisheries, tourism and recreation, and shoreline protection (Burke and Maidens 2004). The Stern Review (2006) estimates that taking no action would mean the loss of 5 percent of global GDP each year, now and forever. Active policy in the form of mitigation is projected to decrease loss to 1 percent.

Mitigation Activities

Society can be involved in climate change mitigation in a number of ways. In a National Academy of Sciences report Fri et al. (2010) suggested a portfolio of approaches:

- Adopt a mechanism for setting an economy-wide GHG emission pricing system.
- Complement this with increases in: a) energy efficiency; b) reduced emissions from energy use in electricity production and transportation including carbon capture and storage plus evolutionary nuclear technologies; and c) retirement, retrofit or replacement of emission-intensive infrastructure.
- Create new technology choices.

Furthermore, they advocated: a) policies that promoted equitable economic outcomes, with attention to disadvantaged populations; b) establishing the U.S. as a leader in GHG reduction; c) encouraging regional and state policy exploration; and d) balancing policy durability and consistency with flexibility and capacity for modification.

Means for Mitigation

The IPCC (2007c) discusses seven key areas for possible mitigation technologies and practices. These are: energy supply, transport, buildings, industry, agriculture, forestry and waste management. In terms of agriculture McCarl and Schneider (2000) list three broad categories of approaches: emissions reductions, sequestration enhancements and product substitution for emissions intensive commodities (fossil fuels and certain building products).

In terms of emissions reductions, Smith et al. (2008) list several strategies: improved agronomic practices, nutrient management, water management (irrigation, drainage), rice management, crop mix change or species introduction (including legumes) and nutrient amendments. For livestock they suggest improved feeding practices, use of specific agents and dietary additives, herd structure/species/animal type change, management changes, and animal breeding, as well as improved manure/biosolid management, storage, handling, anaerobic digestion and more efficient use of manure as a nutrient source.

In terms of carbon sequestration, Smith et al. (2008) list tillage/residue management, agroforestry, set-aside, land-use change, grazing land management, pasture improvement, grazing intensity, fire management, species introduction (including legumes), management of organic soils, avoiding drainage of wetlands, restoration of degraded lands, erosion control, organic amendments, and nutrient amendments. On the forestry side Nabuurs et al. (2007) mention: maintaining or increasing the forest area through reduction of deforestation and degradation plus through afforestation/reforestation; maintaining or increasing the stand-level carbon density through the reduction of forest degradation and via planting, site preparation, tree improvement, fertilization, uneven-aged stand management and other silviculture techniques; maintaining or increasing the landscape-level carbon density using forest conservation, longer forest rotations, fire management, and protection against insects; and increasing off-site carbon stocks in wood products.

In terms of substitute products Smith et al. (2008) list bioenergy feedstocks involving energy crops and residues, while Nabuurs et al. (2007) mention enhancing product and fuel substitution using forest-derived biomass to substitute for fossil fuels or products with high fossil fuel footprints.

One hybrid mitigation method involves pyrolysis and biochar application. McCarl et al. (2009) investigated this prospect and found that it would provide fossil fuel replacement, added sequestration and reduction of fertilizer and water application related emissions. However, they also found that it requires high GHG or fossil fuel prices to be competitive.

Encouraging Mitigation

From an economist's perspective, GHG emissions are an externality in that they are not deliberate acts on behalf of the emitters. Rather, they are production and consumption byproducts that are not really factored into decision-makers' choices. When such emissions cause harm to others the standard economic recommendation is to charge the emitter for the harm, which in turn will influence decisions and choices made by those responsible for the emissions (Arrow 1969; Cornes and Sandler 1986; Cheung 1973). Generally under such circumstances policy approaches include regulations, taxes/subsidies or tradable permits. Their basic characteristics are as follows:

- **Regulation or command-and-control** involves GHG emissions limits that strike a total balance between the damages from emissions and the cost of reducing emissions, including the value of foregone production. In order to set an appropriate limit there is a need for thorough knowledge of production and the damages caused, plus an ability to assure compliance. This may lead to some emitters having substantially higher compliance costs than others.

- **Emissions taxation/reduction subsidization** involves imposing a cost on the volume of emissions or providing a subsidy for emissions reductions. The appropriate tax/subsidy level should represent the damages resulting from additional emissions. The tax alters production decisions and incentivizes emission reductions. Under the tax it is uncertain how far emissions will be reduced as it depends on the nature and amount of the tax (Mankiw 2007).
- **Tradable emission permits** involve a government-imposed limit on GHG emissions in the form of permits for emissions, but the system allows the permits to be traded. It permits emitters that would incur high costs in reducing emissions to buy permits from those who can reduce emissions at a lower cost, lowering the overall cost of emissions (Montgomery 1972; Nordhaus 2008). The Kyoto Protocol advocates such a scheme and spawned the multisector, multi-country European Union Emissions Trading Scheme (EUETS) in 2005. Here total emissions are controlled, but there is substantial price uncertainty. Also, under a tradable permit system producers or firms that are low-cost emission reducers are likely to trade with higher cost emitters, reducing the overall program cost. There are alternatives for the initial allocation of permits with grandfathering and auctioning commonly discussed in this context (Crampton and Kerr 1999; Jensen and Rasmussen 2000).

Generally speaking, economists prefer market-based trading or taxation methods to emissions limits. Through such mechanisms, policy-makers internalize the externality, providing emitters with incentives to reduce emissions and innovate in developing new, lower emitting technologies (Nordhaus 2008).

Difficulties with Policy Design

There are a few basic economic difficulties with the design of policy. Some of these are from the economic literature and other elements that arise from international experience, such as the discussions surrounding the Kyoto Protocol. We discuss these as permanence, additionality, leakage and uncertainty, plus transactions costs and other concerns.

- **Permanence.** One big concern under the Kyoto Protocol is whether emissions reductions are permanent once an emissions reduction activity is undertaken. Certain activities, particularly sequestration activities such as tillage change or afforestation, only provide offsets for a limited time until the land reaches a new equilibrium or the trees mature (West and Six 2007). In addition, the stored carbon is in a volatile form that is released upon practice reduction. This has led to a resistance to biological sequestration and raises the possibility of a price discount for such activities, as discussed in Kim et al. (2008).

- **Additionality.** When one is paid for the use of a practice it is desirable for cost efficiency that this payment stimulates additional GHG emission reductions or additional sequestration. On the other hand, actors who adopt strategies in advance of a mitigation program generally have an interest in being paid for their actions. For example, many U.S. commodity associations have staked out the position that all practices will be paid for regardless of start time. A lot of policy design discussion has focused on additionality (Murray et al. 2007).
- **Leakage.** Another mitigation challenge is leakage. When an emission-reducing action is undertaken in one region it may reduce commodity production, thereby raising prices, and this may be offset by increased production and emissions elsewhere (Searchinger et al. 2008; Fargione et al. 2008). Murray et al. (2004) developed a discounting approach to reducing claimable offsets, but the calculation and scope are difficult because production reactions may be widely spread across the world.
- **Uncertainty.** Agricultural production is highly uncertain and consequently GHG emission reduction and sequestration will also be uncertain. Kim and McCarl (2009) discuss possible discounts as well as the uncertainty-reducing value of widely spread portfolios.
- **Transactions costs.** When GHG offsets are sold the transactions will typically involve cost. Stavins (1995) indicates that for effective policy, transaction costs need to be small. However, enforcement, measurement, monitoring and other costs can mount up (Kim 2011).
- **Other difficulties.** Experience shows that the appropriate design of carbon trading systems can be challenging. Newell and Stavins (2003) indicated that market-based mechanisms are at their best when the cost of emissions abatement differs across emitters. A lack of market power is also needed (Hahn 1984). Also, carbon trading schemes easily adapt to inflation but are not so responsive to technological growth (Stavins and Whitehead 1992).

Who Undertakes Mitigation?

Mitigation actions can be undertaken by both public and private parties. Given emission offset price signals, there will be private industry responses. However, some mitigation activities, such as capturing carbon emissions from power plants, require substantial investments which once developed are likely to spread widely throughout the industry. In those cases the industry cannot privately capture the full benefits of their investment, and as a consequence they are likely to underinvest. Similarly, new mitigation strategies may need to be brought to the attention of emitters. Collectively these imply a public-sector role in the development and dissemination of mitigation technologies.

Also, an atmosphere with reduced GHG concentrations has public good characteristics. However, public goods are subject to free-riding and a lack of recovery of benefits by the parties doing the actual mitigation. This means there will likely be underinvestment, and there may be the need to foster innovation through government means such as research and development subsidies or publically supported research and extension.

Co-benefits and Trade-offs of Mitigation Practices

Many mitigation alternatives not only reduce GHG emissions but also bring other positive environmental or economic results. Advocates of particular mitigation strategies often emphasise their co-benefits. Agricultural tillage changes are often linked to improved water quality, as are fertilization changes and manure management. Wind energy projects reduce the need for coal fired generation with accompanying reductions in mercury, sulfur and ozone-related emissions.

However, to appropriately consider co-benefits, multi strategy trade-offs need to be investigated since the use of some practices may reduce the effectiveness of others with a loss of their co-benefits (Elbakidze and McCarl 2007). Quantification in such a case is difficult.

Adaptation

Adaptation to climate change refers to purposeful actions undertaken to reduce the detrimental effects of climate change and/or exploit beneficial activities. Some degree of adaptation is virtually inevitable. Figure 4.3 above shows alternative colored lines that represent the consequences of alternative mitigation choices. Note that even under the most stringent of those alternatives there is still substantial warming by 2050. Thus we will need to adapt (IPCC 2007c; Rose and McCarl 2008).

Adaptation Means

There are many available agricultural adaptation options including changes in crop mix, adjustment in land use, development of irrigation systems, introduction of new heat- and drought-resistant plant species, use of better fertilization techniques, and changes in planting/harvesting dates.

Public and Private Roles

Private and public decision makers can be involved. Private parties will undertake such adaptations when they are to their benefit. For example, Figure 4.5 shows the observed northwestern shift in corn and wheat production over the last sixty years as the climate has warmed.

Adaptation has been found to have positive impact on the welfare of farmers and societies. However, a big challenge is the cost barrier. The current need for adaptation in agriculture is estimated to be between $9 billion

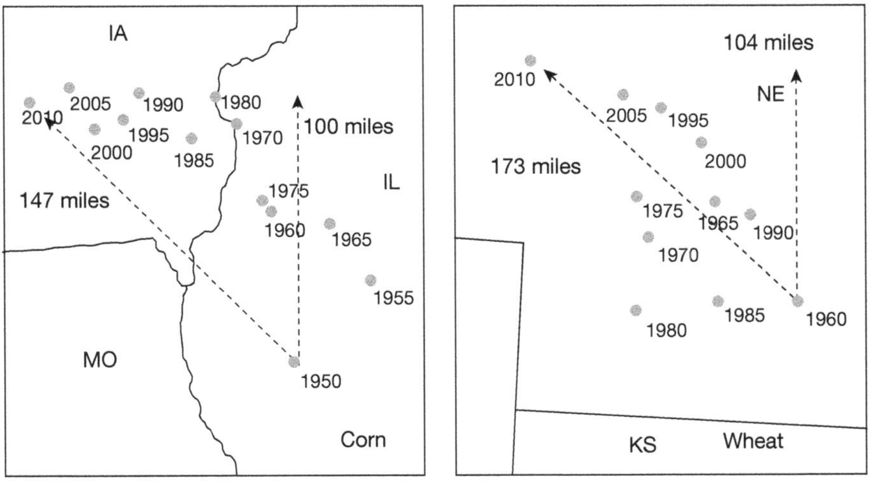

Figure 4.5 Shifts in the Geographic Center of Production for Two Crops (weighted centroids).
Source: Attavanich et al. (2011).

and $40 billion per annum (World Bank 2012), but in 2011 only $244 million in total was distributed for adaptation purposes (FAO 2011). Therefore, adaptation efforts may be perceived as underprovided public good, and both local and public groups might need to be involved in facilitating adaptation actions.

Public action in this area may be of several types. Public adaptation can also be involved, including actions such as: a) research and development on options b) technology dissemination c) infrastructure construction and d) provision of subsidies supporting private action providing possibilities, information and cost relief. Research money would facilitate the development of crop species which are drought- and heat-tolerant and also resistant to pests and fungus (which are more prevalent at higher temperatures). In terms of livestock production, research might find new ways to reduce risks from diseases or develop more heat-tolerant animals. Infrastructure might involve development of transportation routes, water management or coastal protection among many other possible actions. Technology dissemination or extension efforts would disseminate knowledge on adaptation needs and approaches. For example, NOAA, the U.S. Geological Survey (USGS) and NASA created a famine early-warning system, allowing farmers to plan adaptation efforts and thereby decrease potential losses from climate extremes. Technology subsidies may also facilitate adaptation.

Observed Adaptation and Implications

Climate change has led to around a one degree centigrade increase in temperature in the last hundred years, with much of this occurring during the last twenty years. Studies have been done to examine the nature of the observed adaptations and the implications for societal actions. A few sample results follow.

- Reilly et al. (2001, 2002) and Attanavich et al. (2013) studied the way that climate change has caused U.S. crop mixes to shift. They found strong northwest moves. This implies a need for new processing capacity and improved transport systems.
- Mu et al. (2013) studied the way that crop land has changed use to become livestock land and stocking rates have changed. They found lower stocking rates in the south but more grazing land. This may necessitate movements in livestock feeding and processing as well as transport systems.
- Zhang et al. (2013) found a greater incidence of Brahman cattle breeds in hotter Texas areas. Seo and Mendelsohn (2008) found altered livestock species mixes. This may imply a need for handling procedures that better accommodate Brahman animals, and for information dissemination on alternative breeds and species.
- Mendelsohn and Dinar (2003) found that in Brazil and India there are smaller observed yield changes under climate change than agronomic results suggest, indicating that adaptation is present and effective. Tubiello et al. (2000) found adaptations in varieties and planting times. Aisabokhae et al. (2011) found that cropping adaptations are quite valuable. Collectively this says that as of now private adaptation is effective, but it also suggests a need for research efforts on crop timing, breeding and other forms of adaptive management.
- Chen and McCarl (2001a) found that climate change increases pesticide costs, suggesting a need for work on improved methods of pest control under altered climatic conditions.

Conclusions

So at the end of all of these findings, let us address the question: what strategies might form part of a near-term policy agenda? Clearly such an agenda will include a mix of information gathering, mitigation and adaptation. More specifically, agencies at all levels can address a number of issues and data needs. In terms of information:

- Review and assemble evidence on how much climate has changed and in what ways.
- Assemble projections for the region on likely climate change—for example, by drawing together IPCC projections.
- Examine historical data on water, crop yields, crop mixes, livestock numbers, pest incidence and other items, analyzing them to discover climate-related patterns.

In terms of adaptation:

- Given information on climate change, yield and pest changes etc., develop recommendations on adjustments in crops that are likely to be necessary in the next ten years, with livestock adaptation information and implications for infrastructure development and information needs.
- Encourage research and development to address critical needs.
- Consider improving infrastructure to improve adaptation and resilience.

In terms of mitigation:

- As growth and reinvestment in industry, housing, energy generation, and transportation fleets proceed, encourage the development of technologies to reduce greenhouse gas emissions.
- Promote energy efficiency, conservation, and the reduction of emissions from emission-intensive locations.
- Consider taking a position of emissions reduction leadership, engaging in emissions pricing and emissions reduction technology promotion.

Note

1 The A1B scenario is one of the emission scenarios presented in the IPCC Special Report on Emissions Scenarios (2000). Its key assumptions are: people pursue personal wealth rather than environmental quality, rapid economic growth happens, population growth is low, new and more efficient technologies are introduced and regional differences in per capita income are substantially reduced.

References

Adams, R.M. and D.E. Peck. 2008. Effects of climate change on water resources. *Choices* 23 (1): 12–14.

Adams, R.M., R.A. Fleming, C.C. Chang, B.A. McCarl, and C. Rosenzweig. 1995. A Reassessment of the Economic Effects of Global Climate Change on US Agriculture. *Climatic Change*. 30: 147–167.

Adams, R.M., B.A. McCarl, K. Segerson, C. Rosenzweig, K.J. Bryant, B.L. Dixon, R. Conner, R.E. Evenson, and D. Ojima. 1999. Economic Effects of Climate Change on U.S. Agriculture, in R.O. Mendelsohn and J.E. Neumann (eds) *The Impact of Climate Change on the United States Economy,* New York, NY: Cambridge University Press.

Adams, R.M., C. Rosenzweig, R.M. Peart, J.T. Richie, B.A. McCarl, J.D. Glyer, R.B. Curry, J.W. Jones, K.J. Boote, and L.H. Allen. 1990. Climate Change and U.S. Agriculture. *Nature* 345: 219–224.

Aisabokhae, R.A., B.A. McCarl, and Y.W. Zhang. 2011. Agricultural Adaptation: Needs, Findings and Effects, in R. Mendelsohn and A. Dinar (eds) *Handbook on Climate Change and Agriculture*, Northampton, MA: Edward Elgar. 327–341.

Almaraz J.J., F. Mabood, X. Zhou, I. Strachan, B. Ma, and D.L. Smith. 2009. Performance of Agricultural Systems under Contrasting Growing Season Conditions in South-western Quebec. *Journal of Agronomy and Crop Science.* 195: 319–327.

Arrow, K. 1969. The Organization of Economic Activity: Issues Pertinent to the Choice of Market versus Nonmarket Allocation. The Analysis and Evaluation of Public Expenditures: The PPB System. Washington, DC: Joint Economic Committee, 91st Congress, 47–64.

Attavanich, W. and B.A. McCarl. 2011. The Effect of Climate Change, CO_2 Fertilization, and Crop Production Technology on Crop Yields and Its Economic Implications on Market Outcomes and Welfare Distribution. Selected Paper Presented at 2011 Agricultural and Applied Economics Association Annual Meeting, Pittsburgh, PA. July 24–26.

Attavanich, W., B.A. McCarl, Z. Ahmedov, S.W. Fuller, and D.V. Vedenov. 2013. Climate Change and Infrastructure: Effects of Climate Change on U.S. Grain Transport. *Nature Climate Change.* 3: 638–643.

Attavanich, W., B.A. McCarl, S.W. Fuller, D.V. Vedenov, and Z. Ahmedov. 2011. The Effect of Climate Change on Transportation Flows and Inland Waterways Due to Climate-Induced Shifts in Crop Production Patterns. Selected Paper Presented at 2011 Agricultural and Applied Economics Association Annual Meeting, Pittsburgh, PA. July 24–26.

Beach, R.H., A.M. Thomson, and B.A. McCarl. 2010. Climate Change Impacts on US Agriculture. Contributed paper at the IATRC Public Trade Policy Research and Analysis Symposium '*Climate Change in World Agriculture: Mitigation, Adaptation, Trade and Food Security*'. Universität Hohenheim, Stuttgart, Germany, June 27–29.

Burke, L. and J. Maidens. 2004. Reefs at Risk in the Caribbean. Washington, D.C. (USA): World Resources Institute. Available at: www.reefbase.org/resource_center/publication/main.aspx?refid=22167 (accessed 28 October 2013).

Butt, T.A., B.A. McCarl, J. Angerer, P.T. Dyke, and J.W. Stuth. 2005. The economic and Food Security Implications of Climate Change in Mali. *Climatic Change.* 68: 355–378.

Cai, Y. and B.A. McCarl. 2009. Climate Change and Texas Water Planning: an Economic Analysis of Inter-basin Water Transfers. Presented at the 2009 AAEA and ACCI Joint Annual Meeting, Milwaukee, Wisconsin, 26–29 July.

Chen, C.C. and B.A. McCarl. 2001a. An investigation of the relationship Between Pesticide Usage and Climate Change. *Climatic Change.* 50: 475–487.

Chen, C.C., D. Gillig, and B.A. McCarl. 2001b. Effects of Climatic Change on a Water Dependent Regional Economy: A Study of the Texas Edwards Aquifer. *Climatic Change.* 49: 397–409.

Chen, C.C., B.A. McCarl, and C.C. Chang. 2012. Climate Change, Sea Level Rise and Rice: Global Market Implications. *Climatic Change.* 110: 543–560.

Chen, X. and M. Khanna. 2013. Food vs. Fuel: The Effect of Biofuel Policies. *American Journal of Agricultural Economics.* 95(2): 289–295.

Cheung, S. 1973. The Fable of the Bees: An Economic Investigation. *Journal of Law and Economics.* XVI: 11–33.

Corcoran, E., C. Nellemann, E. Baker, R. Bos, D. Osborn, and H. Savelli (eds). 2010. Sick Water? The Central Role of Wastewater Management in Sustainable

Development. A Rapid Response Assessment. United Nations Environment Programme, UN-HABITAT, GRID-Arendal. www.grida.no.

Cornes, R. and T. Sandler. 1986. *The Theory of Externalities, Public Goods, and Club Goods*. Cambridge: Cambridge University Press.

Crampton, P. and S. Kerr. 1999. The Distributional Effects of Carbon Regulation: Why Auctioned Permits are Attractive and Feasible, in T. Sterner (ed.), *The Market and the Environment: The effectiveness of Market-Based Policy Instruments for Environmental Reform*. Cheltenham: Edward Elgar.

Dasgupta, S., B. Laplante, C. Meisner, D. Wheeler, and J. Yan. 2009. The Impact of Sea Level Rise on Developing Countries: A Comparative Analysis. *Climatic Change*. 93: 379–388.

Diffenbaugh, N.S., C.H. Krupke, M.A. White, and C.E. Alexander. 2008. Global Warming Presents New Challenges for Maize Pest Management. *Environmental Research Letters*. 3 044007 doi:10.1088/1748-9326/3/4/044007

Downing, T.E. 1991. Vulnerability to Hunger in Africa: A Climate Change Perspective. *Global Environmental Change*. 1: 365–380.

Elbakidze, L. and B.A. McCarl. 2007. Sequestration Offsets versus Direct Emission Reductions: Consideration of Environmental Co-effects. *Ecological Economics*. 60(3): 564–571.

FAO (Food and Agriculture Organization of the United Nations). 2011. Global Action on Climate Change in Agriculture: Linkages to Food Security, Markets and Trade Policies in Developing Countries. By Aziz Elbehri, Alexandre Genest and Mary Burfisher. Trade and Markets Division. Rome: FAO.

Fargione, J., J. Hill, D. Tilman, S. Polasky, and P. Hawthorne. 2008. Land Clearing and the Biofuel Carbon Debt. *Science*. 7 February.

Fri, R., M. Brown, D. Arent, A. Carlson, M. Carter, L. Clarke, F. de la Chesnaye, G. Eads, G. Giuliano, A. Hoffman, R.O. Keohane, L. Lutzenhiser, B.A. McCarl, M.C. McFarland, M.D. Nichols, E.S. Rubin, T. Tietenberg, J. Trainham, L. Geller, A. Crane, T. Menzies, and S. Freeland. 2010. *America's Climate Choices Limiting the Magnitude of Future Climate Change*, forthcoming National Academy of Sciences Report.

Gleick, P.H., D.B. Adams, A.J. Busalacchi Jr., T. Engman, K.D. Frederick, A.O. Georgakakos, B.P. Hayden, K.L. Jacobs, J.L. Meyer, M.J. Sale, J.C. Schaake, S.S. Seacrest, R. Kuzelka, and E.Z. Stakhiv. 2000. *Water: The Potential Consequences of Climate Variability and Change for the Water Resources of the United States*. Online. Available at: www. gcrio. org/NationalAssessment/water/water. pdf (accessed 19 January 2010).

Godfray, H., J. Charles, J.R. Beddington, I.R. Crute, L. Haddad, D. Lawrence, J.F. Muir, J. Pretty, S. Robinson, S.M. Thomas, and C. Toulmin. 2010. Food Security: the Challenge of Feeding 9 Billion People. *Science*. 327: 812–818.

Hahn, G.L. 1995. Global Warming and Potential Impacts on Cattle and Swine in Tropical and Temperate Areas. *Proceeding of First Brazilian Conference on Biometeorology*. 136–173.

—— 2000. Potential Consequences of Climate Change on Ruminant Livestock Production. Draft USDA/ARS report. Nebraska: U.S. Meat Animal Research Center Clay Center.

Hahn, G.L. and T.L. Mader. 1997. Heat Waves in Relation to Thermoregulation, Feeding Behavior and Mortality of Feedlot Cattle. *Proceedings 5th International*

Livestock Environment Symposium. St. Joseph, MI: American Society of Agricultural Engineers. 563–571.

Hahn, G.L., T.L. Mader, D.E. Spiers, J.B. Gaughan, J.A. Nienaber, R. Eigenberg, T. Brown-Brandl, Q. Hu, D. Griffin, L. Hungerford, A.M. Parkhurst, M. Leonard, W. Adams, and L. Adams. 2001. Heat Wave Impacts on Feedlot Cattle: Considerations for Improved Environmental Management. *Proceedings 6th International Livestock Environmental Symposium*. St. Joseph, MI: American Society of Agricultural Engineers. 129–130.

Hahn, R.W. 1984. Market power and transferable property rights. *The Quarterly Journal of Economics*. 99(4): 753–765.

Hanna, E., F.J. Navarro, F. Pattyn, C.M. Domingues, X. Fettweis, E.R. Ivins, R.J. Nicholls, C. Ritz, B. Smith, S. Tulaczyk, P.L. Whitehouse, and H.J. Zwally. 2013. Ice-sheet Mass Balance and Climate Change. *Nature*. 498: 51–59. doi:10.1038/nature12238.

Houghton, J., L. Meira Filho, B. Callander, N. Harris, A. Kattenberg, and K. Maskell (eds). 1996. *Climate Change 1995: The Science of Climate Change*. Cambridge, UK: Cambridge University Press.

Ifejika Speranza, C., B. Kiteme, and U. Wiesmann. 2008. Droughts and famines: the Underlying Factors and the Casual Links among Agro-pastoral Households in semi-arid Makueni District, Kenya. *Global Environmental Change* 18(1): 220–233.

—1995. *Second Assessment Climate Change 1995. A report of the Intergovernmental Panel on Climate Change*. Online. Available at: www.ipcc.ch/pdf/climate-changes-1995/ipcc-2nd-assessment/2nd-assessment-en.pdf (accessed 1 June 2013).

—— 2000. *Special Report on Emissions Scenarios. Summary for Policymakers. Emissions Scenarios: A Special Report of IPCC Working Group III*. Online. Available at: www.ipcc.ch/pdf/special-reports/spm/sres-en.pdf (accessed 28 October 2013).

—— 2007a. *Climate Change 2007: The Physical Science Basis – Contribution of Working Group I to the Fourth Assessment Report of the Intergovernmental Panel on Climate Change*. Cambridge UK: Cambridge University Press.

—— 2007b. *Climate Change 2007: Impacts, Adaptation and Vulnerability – Contribution of Working Group II to the Fourth Assessment Report of the Intergovernmental Panel on Climate Change*. Cambridge UK: Cambridge University Press.

—— 2007c. *Climate Change 2007: Mitigation of Climate Change – Contribution of Working Group III to the Fourth Assessment Report of the Intergovernmental Panel on Climate Change*. Cambridge UK: Cambridge University Press.

—— *Climate change and Water 2008: Technical Paper VI*.

—— Summary for Policymakers, in O. Edenhofer, R. Pichs-Madruga, Y. Sokona, K. Seyboth, P. Matschoss, S. Kadner, T. Zwickel, P. Eickemeier, G. Hansen, S. Schlömer, and C. von Stechow (eds) *IPCC Special Report on Renewable Energy Sources and Climate Change Mitigation*, Cambridge UK: Cambridge University Press.

—— *Managing the Risks of Extreme Events and Disasters to Advance Climate Change Adaptation. A Special Report of Working Groups I and II of the Intergovernmental Panel on Climate Change* [C.B. Field, V. Barros, T.F. Stocker, D. Qin, D.J. Dokken, K.L. Ebi, M.D. Mastrandrea, K.J. Mach, G.-K. Plattner,

S.K. Allen, M. Tignor, and P.M. Midgley (eds)]. Cambridge UK: Cambridge University Press, p. 582.

Jensen, J. and T. Rasmussen. 2000. Allocation of CO_2 Permits: A General Equilibrium Analysis of Policy Instruments. *Journal of Environmental Economics and Management.* 40(2): 111–136.

Kim, M.K., and B.A. McCarl. 2009. Uncertainty Discounting for Land-Based Carbon Sequestration. *Journal of Agricultural and Applied Economics.* 41(1): 1–11.

Kim, M.K., B.A. McCarl, and B.C. Murray. 2008. Permanence Discounting for Land-Based Carbon Sequestration. *Ecological Economics.* 64(4): 763–769.

Kim, S.W. 2011. The Effect of Transactions Costs on GHG Emission Mitigation for Agriculture and Forestry. Ph.D. Dissertation, Texas A&M University.

Knutti, R. and J. Sedláček. 2012. Robustness and Uncertainties in the New CMIP5 Climate Model Projections. *Nature Climate Change* 1716. Online: 28 October 2012. DOI: 10.1038/NCLIMATE1716

Lobell, D.B., M.B. Burke, C. Tebaldi, M.D. Mastrandrea, W.P. Falcon, and R.L. Naylor. 2008. Prioritizing Climate Change Adaptation Needs for Food Security in 2030. *Science.* 319: 607–610.

Logan, J.A. and J.A. Powell. 2013. Ecological Consequences of Climate Change Altered Forest Insect Disturbance Regimes, in F.H. Wagner (ed.) *Climate change in western North America: evidence and environmental effects.* Allen Press (in press). Online. Available at: www.usu.edu/beetle/documents/Logan-Powell2005.pdf (accessed 31 May, 2013).

Long, S.P., E.A. Ainsworth, A.D. Leakey, J. Nösberger, and D.R. Ort. 2006. Food for Thought: Lower-Than-Expected Crop Yield Stimulation with Rising CO_2 Concentrations. *Science* 312 (5782): 1918–1921.

Mader, T.L., K.L. Frank, J.A. Harrington Jr., G.L. Hahn, and J.A. Nienaber. 2009. Potential Climate Change Effects on Warm-season Livestock Production in the Great Plains. *Climatic Change.* DOI 10. 1007/s10584–009–9615–1.

Mankiw, N.G. 2007. One Answer to Global Warming: A New Tax. *The New York Times, 16.*

McCarl, B.A. 2011. Vulnerability of Texas Agriculture to Climate Change, in J. Schmandt, J. Clarkson, and G.R. North (eds) *Impact of Global Warming on Texas*, Austin, TX: University of Texas Press, ISBN: 978–0–292–72330–6.

McCarl, B.A., C. Peacocke, R. Chrisman, C.C. Kung, and D. Ronald. 2009. Economics of Biochar Production, Utilization, and GHG Offsets, in J. Lehmann and S. Joseph (eds) *Biochar for Environmental Management: Science and Technology,* pp. 341–357. London UK: Earthscan Publisher.

McCarl, B.A. and U.A. Schneider. 2000. US Agriculture's Role in a Greenhouse Gas Emission Mitigation World: An Economic Perspective. *Review of Agricultural Economics.* 22, 134–159.

—— 2001. Greenhouse Gas Mitigation in U.S. Agriculture and Forestry. *Science.* 294: 2481–2482.

Mendelsohn, R. and A. Dinar. 2003. Climate, Water, and Agriculture. *Land Economics.* 79(3): 328–341.

Montgomery, W. 1972. Markets in Licenses and Efficient Pollution Control Programs. *Journal of Economic Theory.* 5: 395–418.

Mu, J. 2012. An Integrated Study of Avian Influenza Impacts and Associated Climate Change Issues. Essay in PhD dissertation. Texas A&M University.

Mu, J.E., B.A. McCarl, and A. Wein. 2013. Adaptation to Climate Change: Changes in Farmland Use and Stocking Rate in the U.S. *Mitigation and Adaptation Strategies for Global Change*. 18(6): 713–730.

Murray, B.C., B.A. McCarl, and H. Lee. 2004. Estimating Leakage from Forest Carbon Sequestration Programs. *Land Economic*. 80(1): 109–124.

Murray, B.C., B.L. Sohngen, and M.T. Ross. 2007. Economic Consequences of Consideration of Permanence, Leakage and Additionality for Soil Carbon Sequestration Projects. *Climatic Change*. 80(1–2): 127–143.

Murray, B.C., B.L. Sohngen, A.J. Sommer, B.M. Depro, K.M. Jones, B.A. McCarl, D. Gillig, B. DeAngelo, and K. Andrasko. 2005. *Greenhouse Gas Mitigation Potential in U.S. Forestry and Agriculture*. EPA-R-05-006. Washington, D.C.: U.S. Environmental Protection Agency, Office of Atmospheric Programs.

Nabuurs, G.J., O. Masera, K. Andrasko, P. Benitez-Ponce, R. Boer, M. Dutschke, E. Elsiddig, J. Ford-Robertson, P. Frumhoff, T. Karjalainen, O. Krankina, W.A. Kurz, M. Matsumoto, W. Oyhantcabal, N.H. Ravindranath, M.J. Sanz Sanchez, and X. Zhang. 2007. Forestry, in B. Metz, O.R. Davidson, P.R. Bosch, R. Dave, and L.A. Meyer (eds) *Climate Change 2007: Mitigation—Contribution of Working Group III to the Fourth Assessment Report of the Intergovernmental Panel on Climate Change*, Cambridge UK: Cambridge University Press.

Newell, R. and R.N. Stavins. 2003. Cost Heterogeneity and the Potential Savings from Market-Based Policies. *Journal of Regulatory Economics*. 23(1): 43–59.

NOAA. 2013. An Interpretation of the Origins of the 2012 Central Great Plains Drought. Assessment Report. NOAA Drought Task Force. Lead: Martin Hoerling, co-leads: Siegfried Schubert & Kingtse Mo. 20 March 2013. Online. Available at: http://drought.gov/media/pgfiles/2012–Drought-Interpretation-final.web-041013_V4.0.pdf (accessed 9 July 2013).

Nordhaus, W.D. 1993. Optimal Greenhouse-Gas Reductions and Tax Policy in the "DICE" Model. *The American Economic Review*. 83: 313–317.

—— 2008. *A Question of Balance: Weighing the Options on Global Warming Policies*. New Haven, CT: Yale University Press.

Olesen, J.E. and M. Bindi. 2002. Consequences of Climate Change for European Agricultural Productivity, Land Use and Policy. *European Journal of Agronomy*. 16: 239–262.

Parry, M., N. Arnell, P. Berry, D. Dodman, S. Fankhauser, C. Hope, S. Kovats, R. Nicholls, D. Satterthwaite, R. Tiffin, and T. Wheeler. 2009. *Assessing the Costs of Adaptation to Climate Change: A Critique of the UNFCCC estimates*. Online. Available at: www.iied.org/pubs/pdfs/11501IIED.pdf (accessed 20 January 2014).

Powell, J.A., and J.A. Logan. 2005. Insect Seasonality: Circle Map Analysis of Temperature-drive Life Cycles. *Theoretical Population Biology*. 67(3): 161–179.

Qureshi, M., M.A. Ejaz, and J.W. Hanjra. 2013. Impact of Water Scarcity in Australia in Global Food Security in an Era of Climate Change. *Food Policy*. 38: 136–145.

Reilly, J.M., F. Tubiello, B.A. McCarl, and J. Melillo. 2001. Climate Change and Agriculture in the United States, in *Climate Change Impacts on the United States: US National Assessment of the Potential Consequences of Climate Variability and Change: Foundation*, Cambridge UK: Cambridge University Press. Chapter 13, pp 379–403.

Reilly. J.M., F. Tubiello, B.A. McCarl, D. Abler, R. Darwin, K. Fuglie, S. Hollinger, C. Izaurralde, S. Jagtap, J. Jones, L. Mearns, D. Ojima, E. Paul, K. Paustian, S. Riha, N. Rosenberg, and C. Rosenzweig. 2002. U.S. Agriculture and Climate Change: New Results. *Climate Change*. 57: 43–67.

Rekacewicz, P. 2006. *Potential Impact of Sea-level Rise on Bangladesh. Collection: COP-5 (the 5th Conference of the Parties) emission graphics*. UNEP/GRID-Arendal. Online. Available at: www.grida.no/graphicslib/detail/potential-impact-of-sea-level-rise-on-bangladesh_a823 (accessed 30 May 2013).

Robinson, A., R. Calov, and A. Ganopolski. 2012. Multistability and Critical Thresholds of the Greenland Ice Sheet. *Nature Climate Change*. Online. Available at: http://www.nature.com/nclimate/journal/v2/n6/full/nclimate1449.html (accessed 11 March 2012). doi: 10.1038/NCLIMATE1449.

Rose, S. and B.A. McCarl. 2008. Greenhouse Gas Emissions, Stabilization and the Inevitability of Adaptation: Challenges for Agriculture. *Choices*. 23(1): 15–18.

Rosenzweig, C., A. Iglesias, X.B. Yang, P.R. Epstein, and E. Chivian. 2000. *Implications of Climate Change for U.S. Agriculture: Extreme Weather Events, Plant Diseases, and Pests*. Cambridge, MA: Center for Health and the Global Environment, Harvard Medical School.

Runyon, J.B., J.L. Butler, M.M. Friggens, S.E. Meyer, and S.E. Sing. 2012. Invasive species and climate change, Chapter 7, in D.M. Finch (ed.) *Climate Change in Grasslands, Shrublands, and Deserts of the Interior American West: a review and needs assessment*. Gen. Tech. Rep. RMRS-GTR-285. Fort Collins, CO: U.S. Department of Agriculture, Forest Service, Rocky Mountain Research Station. pp. 97–115.

Schmitz, H. and H. Shapiro. 2012. The Future of Chocolate. *Scientific American*. 306: 60–65. Online. Available at: http://www.scientificamerican.com/article/the-future-of-chocolate/ (accessed 17 January 2012). doi:10.1038/scientificamerican0212–60.

Seager, R., M. Ting, I.M. Held, Y. Kushnir, J. Lu, G. Vecchi, H.P. Huang, N. Harnik, A. Leetmaa, N.C. Lau, C. Li, J. Velez, and N. Naik. 2007. Model Projections of an Imminent Transition to a More Arid Climate in Southwestern North America. *Science*. 316: 1181–1184.

Seager, R., M. Ting, C. Li, N. Naik, B. Cook, J. Nakamura, and H. Liu. 2013. Projections of Declining Surface-water Availability for Southwestern United States. *Nature Climatic Change*. 3: 482–486. doi:10.1038/nclimate1787.

Searchinger, T., R. Heimlich, R.A. Houghton, F. Dong, A. Elobeid, J. Fabiosa, S. Tokgoz, D. Hayes, and T.H. Yu. 2008. Use of U.S. Croplands for Biofuels Increases Greenhouse Gases through Emissions from Land-Use Change. *Science*. 319 (5867): 1238–1240.

Seo, S.N. and R. Mendelsohn. 2008. Measuring Impacts and Adaptations to Climate Change: a Structural Ricardian Model of African Livestock Management. *Agricultural Economics*. 38(2): 151–165.

Shakhramanyan, N.G., U.A. Schneider, and B.A. McCarl. 2013. US Agricultural Sector Analysis on Pesticide Externalities – The Impact of Climate Change and a Pigovian tax. *Climatic Change*. 117: 711–723.

Smith, P., D. Martino, Z. Cai, D. Gwary, H.H. Janzen, P. Kumar, B.A. McCarl, S. Ogle, F. O'Mara, C. Rice, R.J. Scholes, O. Sirotenko, M. Howden, T. McAllister, G. Pan, V. Romanenkov, U.A. Schneider, S. Towprayoon, M. Wattenbach, and

J.U. Smith. 2008. Greenhouse Gas Mitigation in Agriculture. *Philosophical Transactions of the Royal Society, B Biological Sciences.* 363 (1492): 789–813.

Sortel, C., I. Ibáñez, D.M. Blumenthal, N.A. Molinari, L.P. Miller, E.D. Grosholz, J.M. Diez, C.M. D'Antonio, J.D. Olden, S.J. Jones, and J.S. Dukes. 2013. Poised to prosper? A cross-system comparison of climate change effects on native and non-native species performance. *Ecology Letters.* 16: 261–270.

Stavins, R.N. 1995. Transaction Costs and Tradable Permits. *Journal of Environmental Economics and Management.* 29: 133–148.

Stavins, R.N. and B.W. Whitehead. 1992. Pollution Charges for Environmental Protection: A Policy Link Between Energy and Environment. *Annual Review of Energy and the Environment.* 17: 187–210.

Stern, N. 2006. *Stern Review on The Economics of Climate Change* (pre-publication edition). Executive Summary. London, UK: HM Treasury.

Texas AgriLife Extension Service. 2012. Historic Drought $2.4 Billion Worse than Expected. By William Pack. *The Houston Chronicle.* [Online] 22nd March. Available at: www.chron.com/news/houston-texas/article/Historic-drought-2-4-billion-worse-than-expected-3428512.php (accessed 9 July 2013).

The Renewable Energy Policy Network for the 21st Century. (2013) *Renewables 2013 Global Status Report*, Paris: REN21 Secretariat.

Tubiello, F.N., M. Donatelli, C. Rosenzweig, and C.O. Stockle. 2000. Effects of Climate Change and Elevated CO_2 on Italian Cropping Systems. *European Journal of Agronomy.* 13: 179–189.

United States Climate Change Science Program (USCCSP). 2008. *The Effects of Climate Change on Agriculture, Land Resources, Water Resources, and Biodiversity in the United States.* Washington, D.C.: U.S. Department of Agriculture.

Verbeek, K. 2003. *Climate Change in the Netherlands.* Online. Available at: www.knmi.nl/voorl/nader/klim/klimaatrapportage (accessed 30 May 2013).

Wang, W.W., S. Park, B.A. McCarl, and S. Amosson. 2013. Economic and Groundwater Use Implications of Climate Change in the Texas Ogallala Aquifer Region. Draft. College Station, TX: Department of Agricultural Economics, Texas A&M University.

West, T.O. and J. Six. 2007. Considering the Influence of Sequestration Duration and Carbon Saturation on Estimates of Soil Carbon Capacity. *Climatic Change.* 80: 25–41.

World Bank. 2012. Turn Down the Heat. Why a 4° Warmer World Must be Avoided. A Report for the World Bank. Potsdam: Institute for Climate Impact Research and Climate Analytics. November 2012. Online. Available at: http://climatechange.worldbank.org/sites/default/files/Turn_Down_the_heat_Why_a_4_degree_centrigrade_warmer_world_must_be_avoided.pdf (accessed 2 June 2013).

Zhang, Y.W., A.D. Hagerman, and B.A. McCarl. 2013. Influence of Climate Factors on Spatial Distribution of Texas Cattle Breeds. *Climatic Change.* 118: 1–13.

5 Socio-Economic Implications of Nuclear Power on Rural Communities

Beth-Anne Schuelke-Leech

Introduction

Nuclear power has been on the cusp of a renaissance for many years, or so it has seemed to many within the industry. The need for nuclear power has seemed virtually inevitable. Nuclear power currently provides about 19 percent of the electricity in the United States (U.S. Energy Information Administration 2012). This is electricity that does not come with the burden of heavy greenhouse gas emissions, since nuclear power has none of the combustion emissions of coal, natural gas, and oil. In addition, nuclear power is capable of meeting base-load power needs cleanly, as it does not suffer from the intermittency problems of solar or wind power (Dell and Rand 2001; Anderson and Leach 2004), which necessitates back-up conventional power. Nuclear power is burdened by the well-publicized risks involved with the use of radioactive fuels,[1] as well as policy and technology decisions that have locked in the way that nuclear power is produced and managed. However, the greatest impediment to a nuclear renaissance may be the fundamental financial costs that slowed the industry to a virtual halt in the 1970s. This chapter examines the nuclear industry in the context of these technological, political, and financial factors and explores the potential risks and benefits of nuclear energy for rural communities. Currently, the impact on rural communities is relatively small. However unlikely, a nuclear renaissance would bring both potential benefits from the reduction of greenhouse gas emissions and increased risks if the problems of nuclear power are not dealt with effectively. The next section presents the historical background of nuclear power. Then the risks and benefits of nuclear power for rural communities are presented and discussed.

Background

Nuclear power was not developed in response to a market need or entrepreneurial ingenuity. The first sustained nuclear reaction took place in 1942 in a squash court at the University of Chicago.[2] The power of nuclear reactions was evident, and the United States military instituted a massive program to harness nuclear energy, first for destructive purposes through the development of nuclear bombs, and later through the development of nuclear power reactors. Initially, reactors were developed and used to power submarines. The adoption of these light water reactors for commercial purposes was both pragmatic and problematic. While it did allow for faster development and deployment of nuclear energy for commercial power generation, it also locked the industry into a particular design that was not necessarily optimal for land-based electrical power plants. Utilities quickly adopted nuclear technology, often in response to incentives from reactor manufacturers and overly optimistic forecasts of construction costs and energy demand. Thus, nuclear power was produced by technological, political, and financial willpower. This history significantly influences the potential implications for its continued use and future.

There are 103 nuclear power plants currently operating in the United States. Approximately one third of these power plants are sited in rural jurisdictions (see Figure 5.1). Though many of the power plants in the United States were originally located in rural areas, substantial population increases in the past fifty years means that most are now located in suburban or urban areas. Since 2000 there has been an average increase of 10.9 percent in the population living within a five-mile radius of nuclear power plants, and a 27.4 percent increase in those living within a fifty-mile radius (Dedman 2011). Thus, any expectation that isolation from urban centers would provide some measure of protection to the population from any nuclear accident is no longer valid.[3]

Communities near nuclear power plants bear the risks of a nuclear accident or radiation leakages. Despite this, there is no empirical evidence that this risk adversely affects the price of property or houses near a nuclear power plant (see, for example, Gamble and Downing 1982; Kiel and McClain 1995; Jackson 2001; Bezdek and Wendling 2006), or that people are reluctant to live near them for this reason alone (Donn 2011), given the population increases around nuclear power plants (Dedman 2011).

While nuclear power provides the potential to transition away from more carbon-intensive energy sources, there are still significant problems with nuclear power that need to be addressed, including safety, management of the fuel cycle, nuclear waste, and economics and cost. Each of these must be sufficiently addressed if nuclear power is to be considered a viable and ongoing (and possibly increasing) component of energy produced.

Challenges

The challenges faced by the nuclear industry fall into three broad categories: safety and waste management, economic and environmental impacts.

Safety and Waste Management

While nuclear power has a safe operating record relative to other conventional energy sources (Deutch, Forsberg et al., 2009), the possibility of a nuclear meltdown and public concerns about exposure to radiation make nuclear power an ongoing safety concern for many people. In addition, examples of safety and environmental concerns, such as the power plant accidents at Three Mile Island in 1979 (Walker 2004), Chernobyl in 1986 (Medvedev 1990; Medvedev 1991; UNSCEAR 2008) and Fukushima-Daiichi in 2011 (ANS 2012), the environmental contamination of land surrounding uranium mining sites and uranium enrichment facilities (Ringholz 2002; Amundson 2004; Malin and Petrzelka 2010), and the collapse of the water tower at the Vermont Yankee Power Plant in August 2007 (Watts 2012), have raised concerns about the abilities of nuclear power plant owners and operators to address safety concerns and act in the best interests of public health and safety.

However, it is impossible to assess the safety of the nuclear industry without some context. All energy production and industrial processes have some risks. The nuclear industry has an impressive record of safe operations, with accident and fatality rates well below comparable activities (see Table 5.1).

Many people have a significant fear of being exposed to radiation, well beyond their likelihood of receiving a significant dose. This is undoubtedly because radiation is so closely tied with nuclear weapons and the effects of atomic bombs. The general population finds it extremely difficult to differentiate between the hazards of nuclear war and those of nuclear power generation. Consequently, there is no way to separate this fear of radiation from nuclear power or nuclear waste. Some exposure to radiation is normal since radioactive materials occur naturally (World Nuclear Association 2013b). For instance, one kilogram of coffee has a radioactivity measure of 1,000 Bq.[4] The average person has a radioactivity of 65 Bq/kg. The average household detector (with americium) has a radioactivity measure of 25 million Bq, the same measure as 1 kilogram of uranium. A radioisotope source used for medical diagnosis has a radioactivity measure of 70 million Bq. Problems generally only occur when exposure is acute or severe.

Radiation is measured by the dosage received. Radiation is measured in either siverts (Sv) in the International System of Units, or rems (rem).[5] In 2011, the worldwide annual exposures to natural background radiation were in the range of 100 to 1300 mrem, with the average being approximately 300 mrem, while the average measurable dose per individual from nuclear power plants was 170 mrem (U.S. Nuclear Regulatory Commission 2011). Thus, for most people, the background radiation that they receive naturally is actually greater than any exposure resulting from living near a nuclear power plant. For individuals undergoing any medical treatment that employs radioactive materials, or for those exposed to naturally occurring radon, their annual dosage can be much higher.

Table 5.1 Safety Record by Industry, 2003–2012

Industry	*Average annual incidence per 100 full-time workers*	*Total incidence per 100 full-time workers*	*Average annual fatalities*	*Total fatalities*
Oil and Gas Extraction	87.8	175.6	17.6	158.0
Crude Petroleum and Natural Gas Extraction	60.6	121.1	12.1	109.0
Coal Mining	151.7	303.3	30.3	273.0
Metal Ore Mining	28.3	56.7	5.7	51.0
Uranium–Radium–Vanadium Ore Mining	0.0	0.0	0.0	0.0
Oil and Gas Pipeline and Related Structures Construction	86.7	173.3	17.3	156.0
Power and Communication Line and Related Structures Construction	155.0	310.0	31.0	279.0
Chemical Manufacturing	123.9	247.8	24.8	223.0
Plastics and Rubber Products Manufacturing	92.2	184.4	18.4	166.0
Rail Transportation	94.4	188.9	18.9	170.0
Pipeline Transportation	14.4	28.9	2.9	26.0
Motor Vehicle Manufacturing	81.1	162.2	16.2	146.0
Electric Power Generation	131.7	263.3	26.3	237.0
Fossil Fuel Electric Power Generation	20.0	40.0	4.0	36.0
Nuclear Electric Power Generation	0.0	0.0	0.0	0.0
Electric Power Transmission and Distribution	77.8	155.6	15.6	140.0
Natural Gas Distribution	18.9	37.8	3.8	34.0
Water, Sewage and Other Systems	18.3	36.7	3.7	33.0

Source: U.S. Department of Labor, Bureau of Labor Statistics, in cooperation with State, New York City, District of Columbia, and Federal agencies, Census of Fatal Occupational Injuries.

However, people's fear of radiation is generally associated with acute, fatal exposure to radiation caused by a catastrophic event, or the increased adverse health problems that result from long-term exposure to non-naturally occurring radiation. Assessing the risk is problematic. In general, there are objective (also known as real or scientific) risks and subjective (or perceived) risks. However, definitions and assessments of risks by scientific experts and risk professions differ from those used by the general public (Freudenburg 1988; Gardner and Gould 1989). The public generally use a broader, more complex definition of risk (Gardner and Gould 1989). Covello (1983) notes that people use two important guidelines or rules in assessing risk: (1) the availability of information about an event and how easy it is to recall; and (2) the representativeness or similarity of events. Covello points out that people tend to evaluate an event as more probable or frequent if they can recall it quickly, and that people assume that similar technologies and events have similar risk levels. This helps to explain why people have a heightened sense of potential risks for nuclear power: the major incidents are easy to recall and there is a perceived representativeness of all nuclear technologies by the explosion and destruction of atomic weapons. It also helps to explain why people underestimate the likelihood of common health risks and deaths (e.g. heart attacks) but overestimate those attributed to events with high media coverage (e.g. homicides, natural disasters, nuclear accidents) (Combs and Slovic 1979; Slovic, Fischhoff, and Lichtenstein 1980). These latter events are also now subject to much greater publicity and speculation as they unfold, with the rise of instant and social media. Thus, people often take note of the original estimates made as an event unfolds, rather than subsequent revised estimates.

Radiation exposure limits were developed based on studies done after atomic bombs were dropped on Hiroshima and Nagasaki in Japan in 1945. Researchers found that exposures of 600 rems or more of radiation were fatal 100 percent of the time, while 450 rems was fatal approximately 50 percent of the time (*MIT Tech Talk* 1994). Annual exposure limits assumed that radiation exposure is cumulative and can have the same adverse health effects as acute exposure (Health Physics Society 2013). Though the actual health effects of exposure will vary depending on occupation, weight, height, and age, currently adults are limited to an annual occupational radiation exposure limit of 5 rems above the 0.3 rems expected from natural sources. Annual limits for the public and specific occupations are set by several federal agencies, including the Department of Energy, the Nuclear Regulatory Commission, and the Environmental Protection Agency, and state agencies (*MIT Tech Talk* 1994).

The effects of a nuclear accident are significant. The nuclear power industry has experienced four significant accidents. The first was in Idaho in 1961, where three workers died after their military research reactor had a critical failure that resulted in a meltdown of the reactor. However, this accident is relatively unknown and it had virtually no impact on the pace of the adoption of nuclear power.[6] The second accident was at the Three Mile Island power plant in 1979 when there was a partial meltdown of the reactor. Fortunately there was no leakage of reactive materials and no loss of life. In 1986, the third major accident happened at the Chernobyl nuclear power plant in Ukraine. The technological failure, coupled with political decisions to deny the accident, had a deeply scaring effect on the public and the industry globally.[7] However, until the Fukushima-Daiichi accident in March 2011, it was possible for many to deny that such an accident was possible with Western nuclear power technologies. The environmental, social, political, and economic effects of such an accident are undeniable. The land around the Chernobyl power plant is still unusable and long-term health effects exist, though they are extremely difficult to measure accurately. However, Fukushima-Daiichi showed the world that no matter how nuclear power technology has performed or been managed in the past, there are always inherent dangers of a meltdown, particularly with light water reactors that require active intervention to ensure that the reactor is properly cooled. If anything interferes with this cooling process, the reactor is vulnerable to a meltdown. The conventional wisdom within the industry is that a meltdown is highly unlikely when the power plant is properly managed. This may be a fair assessment, but as the accidents have shown, meltdowns are possible, particularly when combined with natural disasters (such as a tsunami or earthquake) or the coupling of technological failure and human error (as with Three Mile Island and Chernobyl).

The implications of a nuclear meltdown are substantial for surrounding rural communities, even if the reactor itself is in an urban location. First and foremost, rural communities necessarily become the initial destination for evacuees if they need to leave the local urban area. In looking to escape exposure to radiation, people near the reactor will necessarily end up in the surrounding areas, even if only temporarily. Evacuation shelters and other facilities are relatively limited in rural communities, and thus it is likely that evacuees would pass through rural areas on their way to larger communities, or else they would need to stay within their own urban jurisdiction. However, any evacuation to rural communities would stress local governments, infrastructure, and communities. This would also be true for the rural communications networks, which are not typically as dense as those in urban settings. Local governments have limited resources and are not equipped for large-scale operations. This need would greatly increase if fallout from a nuclear accident affected rural lands directly through radiation deposits, as occurred in the Fukushima prefecture in Japan. Though urban governments and state and federal agencies would undoubtedly try to help evacuees by providing facilities and other resources, coordinating these resources in disasters and getting them to where they are needed can be difficult, as evidenced by the events around the Fukushima-Daiichi accident and Hurricane Katrina.

In the event of an accident, clean-up and decontamination will necessarily involve the surrounding rural areas, whether the actual reactor is in an urban or rural setting. Rural areas are likely to be the staging ground for clean-up, receiving materials as they are being cleaned and providing replacement soil. Clean-up challenges are substantial, as the Fukushima-Daiichi accident has shown. Unlike the Chernobyl accident, where many of the details were hidden from the media and the public, Fukushima-Daiichi has been well-publicized. Large volumes of contaminated water, debris, soil, secondary wastes, and spent fuel must be dealt with during the clean-up process (ANS 2012). There is no question about the fact that any surrounding rural areas would be significantly involved and affected by an accident.

Large-scale evacuations and disaster relief, such as occurred during Hurricane Katrina and Fukushima-Daiichi, necessarily expose the disparities that exist among people. Leaving a disaster zone requires the financial and social capital to be able to move to a new location. Typically, those who do not have the necessary resources are the poor and often minority populations (Berube, Deakin, and Raphael 2006). It is important to note that in disaster situations people often lose the social capital and networks that allow them to navigate through their daily lives. Rural communities are often close-knit, with neighbors providing physical and emotional support to each other. During large-scale evacuations and displacements, this local community support is disrupted as residents leave and are then compelled to develop new support networks. The difficulty in this situation can be exacerbated by the fact that rural residents may be forced into non-rural settings during evacuations and afterwards.

Safety in the nuclear industry is closely tied to waste management, since both are really about the health and environmental risks of radioactive materials. The strategy for dealing with nuclear waste has consisted of long-term storage of highly radioactive materials in dry cask systems, typically in an isolated rural location. Yucca Mountain, Nevada was the proposed site for a national nuclear waste repository (Walker 2009). Protests and local opposition to the development resulted in Yucca Mountain ceasing to be the proposed site (Riddel, Dwyer, and Shaw 2003). Currently, no suitable alternative site has been agreed upon and waste continues to be "temporarily" stored on-site at power plant facilities (World Nuclear News 2009).

Waste is really better considered as depleted fuel where a portion of the uranium has been used, meaning that it is no longer suitable as fuel for nuclear reactors (of the current design); this is the reason that waste is also referred to as spent fuel. Spent fuel can be reprocessed and reused, as is done in France (Hecht 1998), eliminating a large component of the waste problem. However, due to concerns about the proliferation of weapons-grade fissile material, reprocessing in the United States is prohibited under restrictions originally put in place by President Jimmy Carter (Mahaffey 2009). Currently spent fuel is stored in spent fuel pools or in dry cask storage, typically at nuclear power plant sites, because the Department of Energy has not yet established a long-term repository (World Nuclear News 2009). This strategy is really evidence of the inability of policymakers and the nuclear industry to deal with the waste issue. The technology exists to reprocess waste. Storage is not the only option, and in fact reprocessing allows more energy to be derived from the fuel before it is expended and thus reduces the amount of waste that ultimately needs to be stored. In addition, there are other types of reactor designs that use nuclear fuel more effectively and completely. The difficulty is that these designs have limited development and use thus far, and are unlikely to be adopted in the foreseeable future. This makes storage the only option.

Rural locations are really the only viable option for storage. Yucca Mountain is no longer the designated location (Walker 2009) and currently there are no tangible plans for a centralized repository for waste (World Nuclear News 2009). Arguably this increases the risks at each power plant. However, one centralized location would substantially increase the likelihood of exposure for the rural area where it is located. The transportation of the waste to a central repository would also tend to disproportionally expose rural areas, as the hazardous material would likely be routed to avoid population centers. This was a central issue in the protests around Yucca Mountain as a nuclear waste repository (Riddel et al. 2003).

Regardless of how the waste is dealt with or the exact location of the reactor, there is no question that rural areas bear a disproportionate burden in the risks of nuclear power relative to the amount of power that they consume, just as they do with most energy sources.

Economics

Though it might be tempting to believe that the environmental movement and the Three Mile Island accident were really the reasons that the utilities stopped ordering and building nuclear reactors, this is not accurate. Fundamentally, it was economics that made nuclear power unattractive and uncompetitive. Cost over-runs and operations delays caused utilities to move away from nuclear power in the 1970s (Campbell 1988).[8] Nuclear power plants have proven to be cost effective and economical over their long lifespan, providing relatively cheap power per unit of energy. However, the initial capital costs make them prohibitively expensive. The Energy Information Agency (U.S. Energy Information Administration 2012) estimates that it would cost approximately $11.7 billion to construct a new 2,200 MW nuclear power plant, equating to approximately $5,335 per kilowatt of capacity. The U.S. federal government allocated $18.5 billion in incentives and loan guarantees for new builds. However, $8.3 billion alone was allocated to the new Vogtle units.[9] Compared to natural gas and coal, it is very difficult to justify allocating so much to nuclear power. Though history has shown that nuclear power plants do yield a positive return on the capital investments made to construct them and they are cost efficient to operate, changes in economic, political, and financial environments have made additional nuclear power capacity difficult to justify.

Since the 1970s, four important trends have impacted the nuclear industry, adversely affecting the economic business case for reactors. The first is the rise of the political ideology that has its foundation in neoliberal economic theories, with a focus on market solutions and limited government. Nuclear power is not simply another energy source from a market perspective, however. Nuclear power does not compete openly. It is regulated by the federal government via the Nuclear Regulatory Commission. From the mining and enrichment of uranium, through the building and licensing of reactors and the transportation and use of nuclear fuel to the storage of the waste, the government closely controls nuclear materials and nuclear power. Utilities cannot experiment or innovate with their nuclear operations in the way that they can with other sources of fuel. In the nuclear industry, government and regulators are ever-present. This puts nuclear power at odds with the current dominant political and economic ideology of free-market competition and limited governmental regulations.

The second major influence on the economics of nuclear power plants is the rise of financialization and the philosophy of maximizing share price and earnings per share in the short-term as the means to maximize shareholder value. This has shifted the economic focus towards financial wealth and short-term metrics, and away from capital wealth and returns from long-term investments. Rather than investing in research and development and long-term capital projects, many firms are using their capital to manipulate stock prices and taking actions to meet short-term financial targets (Lazonick 2008). This makes it much more difficult for utilities to make long-term investments, such as building nuclear power plants, even if such investments will ultimately prove to be profitable.

The third factor is the rise of the environmental movement and citizen activism in the licensing and operation of nuclear reactors that began in the 1970s and has continued to the present, often resulting in delays to nuclear projects. Changing regulatory requirements to meet environmental and safety concerns have forced utilities to alter designs during construction, often after critical decisions had already been made. These delays have been costly and were often unanticipated by the utility companies. Many utilities ran into financial difficulties when their plants were delayed because under public utility regulations, utilities could not recover costs by passing them along to consumers through higher prices until the delayed plants were actually operating (McCaffrey 1990). With the high interest rates of the 1970s and 1980s, utilities were placed under increasing financial stress as they tried to finance construction costs and complete their projects.

The fourth factor, changing energy demands, also contributed to this financial stress because forecasted revenues that formed the foundation of the economic justification for new nuclear power plants were not realized.

The oil shocks of the early 1970s led to reduced energy consumption and greater conservation. This meant that the anticipated growing needs for energy did not materialize. Utilities simply could not justify investing in large-scale increased capacity the same way that they had been able to in the 1960s.

These trends have made the economic justification for nuclear power plants extremely dubious. Nuclear power plants are very complicated—far more so than conventional fossil fuel plants. A typical fossil fuel plant has approximately four thousand valves and five thousand pipe supports. A nuclear plant might have ten times as many valves, and tens of thousands of pipe supports (McCaffrey 1990). Many utilities and contractors had difficulty managing the complexity and uncertainty of nuclear power plants, making the financial situation worse. These factors also meant that nuclear power never yielded the expected economic benefits of standardized reactor designs, mass manufacturing, and the development of experience and best-practices. Each reactor is essentially a custom design and build. This means that there is no way to spread costs over multiple builds. Given the current economic realities, few utilities are willing to voluntarily undertake the financial risks associated with nuclear power again.

Environmental Impacts

Nowadays there are substantial economic and environmental concerns over the construction and operation of nuclear power plants. The operating life of many nuclear facilities has been extended for up to eighty years (World Nuclear Association 2013a). However, eventually these sites will need environmental remediation. There have been several successful examples of nuclear facilities being decommissioned and demolished and the surrounding lands being restored, including the Big Rock Nuclear Power Plant in Charlevoix, Michigan, the Connecticut Yankee in Haddam Neck, Connecticut, the Maine Yankee in Wiscasset, Maine, the Yankee Rowe in Rowe, Massachusetts, and the Trojan Nuclear Power Plant near Rainier, Oregon. Unfortunately, much of environmental history shows that corporations do not necessarily do an adequate job of protecting the environment or surrounding communities, especially when they are leaving a site.

Climate change and increasing pressures on water resources may also have a significant impact on nuclear energy. The production of electricity through nuclear power plants requires a substantial amount of water for their operations (Kirkwood 1982; Feeley III, Skone et al. 2008; Fthenakis and Kim 2010). This water comes either from nearby natural water sources such as lakes, rivers, or oceans, or from man-made reservoirs. With the expected increase in temperatures caused by global warming, two problems arise. First, the temperatures of these cooling water sources are expected to rise, meaning that less waste heat from energy production can be absorbed by the water (Sovacool and Sovacool 2009). Second, as global temperatures increase, greater evaporation will mean that in many of these water sources there will be less water available for cooling (Chandel, Pratson, et al. 2011). Both of these problems mean that nuclear power plants will have to reduce their water consumption, likely meaning reduced electrical output (unless power plant operators can figure out how to reduce their demands for water). The same considerations apply to fossil fuel power plants (mainly coal and natural gas). These issues around water lead to further questions about the rights to water. Water is essential for electricity production. It is also needed for food production—a vital rural activity. If there are new restrictions on water availability, then there will need to be widespread agreements about priorities for the water.

Food production is also at greater risk around nuclear power plants since much of the rural land around nuclear sites is used for agricultural purposes. On average, 38 percent of the land in the counties in which nuclear power plants are operating is used for farms and agricultural production (author calculations from U.S. Census, 2013). Concerns over the contamination and radiation of water resources and food supplies rose substantially after the Fukushima-Daiichi accident (ASME 2012; Klein and Corradini 2012). Even if the actual dose and risk from radiation is insignificant, if the public refuses to purchase agricultural products from a particular district, this could have substantial adverse effects on the regional economy and the inhabitants of rural areas. There is no indication that the public changes its consumption patterns based on food being grown in proximity to nuclear reactors, but this remains a possibility.

Most of the fuel for nuclear power plants is uranium, which is mined and then processed and enriched. Mining is almost exclusively concentrated in rural areas in the Western states, including Montana, Wyoming, Colorado, Utah, and New Mexico. In general, uranium mining has proven to be both individually and environmentally harmful (Ringholz 2002; Amundson 2004; Malin and Petrzelka 2010). Rural communities have much greater exposure than urban communities to this environmental pollution through potential ground water contamination, waste tailings, or long-term exposure. In addition, the raw fuel, enriched uranium and waste are transported throughout rural areas as they are moved across the country (Walker 2009). This places the burdens of environmental problems from mining on the rural areas in which they occur.

Communities and individuals in the areas surrounding nuclear power plants bear potential risks without necessarily having any real democratic participation in the issues. Nuclear power plants are licensed and regulated by the federal government via the Nuclear Regulatory Commission (NRC). These regulations supersede state and local laws, which generally have little influence over nuclear power plants. Though environmental assessments and monitoring are required, it is unclear that these take into account the social and economic impacts of proximity to a nuclear power plant. However, the cases of the Shoreham Nuclear Power Plant on Long Island,

New York, which was constructed, licensed, and then demolished without ever operating (McCaffrey 2004), and the Vermont Yankee Nuclear Power Plant in Vernon, Vermont (Watts 2012), which continues to struggle with active state and local attempts to shut the plant down, illustrate how local communities and stakeholders can have a significant influence on the operation and success of nuclear power.

Benefits

Nuclear power has some significant benefits that should not be discounted.

Economic

One of the significant benefits of nuclear power for rural communities is the jobs and economic development that comes from the construction and operation of a power plant. Constructing nuclear power plants requires thousands of workers, especially skilled trades, construction workers and engineers. Operating nuclear power plants also requires hundreds, and sometimes thousands, of highly skilled workers.

Nuclear power plants can take anywhere from five to ten years to build (occasionally longer), and they have operating lives extending to eighty years. Thus, in addition to the immediate economic benefit, the siting of nuclear power plants can provide stable employment for upcoming generations. One of the ongoing challenges for rural communities, particularly those far from metropolitan areas, is providing well-paying, consistent employment to their residents. Young people are often faced with the choice of leaving their rural home to find employment or else staying in the local area without access to employment opportunities. Nuclear power plants can provide an avenue to well-paying, stable employment for educated workers, without forcing them to sacrifice social networks and connections.

Furthermore, the critical mass of people employed and benefiting economically from the power plant may actually promote the development of technological and physical infrastructure, both by providing a stable tax base for local governments and by providing the impetus for sufficient demand for those services by the population.

In addition to the employment benefits, once a plant is operating nuclear power can provide relatively cheap power for many years. The average utilization of nuclear power plants (known as the "capacity factor") increased from 56 percent in 1980 to over 90 percent in 2009 because of ongoing improvements in operating efficiency. This allows utilities to employ nuclear power more consistently to keep their energy supply costs relatively low and stable. This is a benefit not only for the utilities, but also for local consumers.[10]

Clean Energy

Nuclear power remains the one power source able to meet base-load energy needs without the direct emission of greenhouse gasses. Ultimately, the justifications for nuclear power come down to whether clean energy is a necessity. Though many environmentalists and advocates for renewables would like to see more renewables (and this is desirable), these kinds of energy sources have difficulty consistently meeting base-load power requirements. Renewables need to be backed up by some source of electricity generation that can operate when the renewable power is not available. This back-up is currently provided largely by various fossil fuels. Though renewables may eliminate the need for the use of some fossil fuel, they do not eliminate it all. In addition, once the cost of the emitted carbon to produce the equipment to harness the renewable energy is considered, there is a long payback period to offset the carbon emissions.

Nuclear power production is also partially dependent on carbon-based energy, as the mining and processing of uranium rely on petroleum and electricity produced by a variety of sources, as does the construction of the power plant, the reactor, and the materials. Thus, the main question is whether the clean production of electricity is needed, and which source of power is going to be used to meet base-load demand, provide back-up for renewables, and supply the massive amounts of energy that will be needed to extract the remaining fossil fuels that are less and less accessible.

While considering the rural implications of using nuclear power, it is also necessary to think about the impacts on rural areas if nuclear power is not used. Unless there is a major shift in policy and energy production, it is likely that the majority of energy will continue to be produced from fossil fuels, the effects of which are talked about elsewhere in this book. At current rates, there is no expectation by scientists and environmentalists that we will be able to level off atmospheric carbon concentrations below 650 parts per million (Anderson and Bows 2010). Though this figure is far from exact, it will likely result in at least a four degree rise in global temperatures. With this increase, it is far more likely that there will be increasing sea levels, flooding coastlines, disruptions in current weather patterns, increased storms and droughts, and changing food production patterns. It is impossible to predict the exact impact that this will have on any given area, but it is likely that as the average global temperature rises and climate change problems become more visible, there will be greater pressures directed towards policymakers to address these challenges. However, as the situation becomes more critical and more urgent, the time available to develop and implement solutions will decrease. Five years ago, many nuclear power experts anticipated an imminent nuclear renaissance (Nuttall 2004; World Nuclear Association 2011). New light water reactor designs have been developed to meet this expected renaissance, particularly Small Modular Reactors (SMR) such as the AP1000, which are considered safer and less costly. With the widespread introduction of advanced drilling technologies such as hydraulic fracturing and horizontal drilling in the United States, the focus is now on the extraction of natural gas and the expected future of abundant, cheap, domestically-produced fossil fuel (see, for example, Kinnaman 2011; Institute for Energy Research 2012; Weber 2012; Kamalick 2013; *The Economist* 2013). However, though natural gas is a cleaner-burning fuel than either coal or oil, it is still a carbon-based fossil fuel and its combustion still produces greenhouse gasses. Renewable energy sources represent less than 10 percent of electricity production (U.S. Energy Information Administration 2012), primarily because of their intermittency in production (Dell and Rand 2001; Anderson and Leach 2004). Thus, renewables are unlikely to reliably and consistently meet all demands for energy. Nuclear power does not emit greenhouse gasses directly. Thus, it provides a means to supply non-carbon base-load power. Nonetheless, there are substantial technical, environmental, and public concerns about nuclear power.

One of the proposed solutions to safety and economic concerns is new reactor designs. There are currently five new reactor designs being developed other than the SMR: (1) gas-cooled reactors; (2) liquid metal reactors; (3) super-critical reactors; (4) molten salt reactors; and (5) sodium-cooled reactors. The Chinese government is currently working on each one of these design options in parallel, and there are other development projects going on around the globe. However, these designs are at various stages of development, they are largely untested, and they are not ready for large-scale commercial deployment. This means that if there is increased pressure to construct and license nuclear reactors in the shorter term, there is less likely to be any real innovative advances in their design and safety. Instead, it is far more likely that the United States will continue to utilize light water reactors, with their inherent need for active management for safety.

Conclusions

Unquestionably, nuclear power is an energy source that has great economic, technological, and political challenges. It has advocates and opponents, many of whom are utterly committed to their feelings about the benefits or evils of the energy source. In the shorter term, it is unlikely that the proportion of electricity produced by nuclear power is likely to change. After the Fukushima-Daiichi accident, the global nuclear power industry underwent a critical review of both the accident and its own practices. There is a clear sense in the nuclear power industry that it is a global industry and that challenges in one country must be addressed by all. Many countries have reassessed their use of nuclear power. The German government announced plans to eliminate all nuclear power plants. The Japanese are still struggling to get their nuclear reactors back online, after they were all shut down for safety reviews after the accident.

Nonetheless, the need for energy is not going away. Individual countries may choose to reject nuclear power as an energy source, but they still require energy. For most countries, this largely leaves fossil fuels as the primary means of production, with all of its attendant challenges. Though it would be preferable if renewables could meet all our energy needs—and certainly there could be much wider use of renewables—they are an intermittent source of power and still need back-up from non-renewable sources, at least until sufficient inexpensive battery storage is available. Thus, for now, base-load power must come from fossil fuels or nuclear power.

Until the challenges of nuclear power are dealt with, however, it is unlikely that there will be widespread acceptance of the desirability of nuclear power, or a willingness to invest in it. The entire lifecycle of nuclear power production, from mining through waste management, must be wisely managed and policies put in place to support the effective and safe use of this technology. For rural communities, as for other communities, the benefits of cleaner base-load power may eventually outweigh the economic challenges. For now, economics alone simply do not favor significant numbers of additional nuclear power plants.

Acknowledgements

The author wishes to thank the two anonymous reviewers for their helpful suggestions and comments.

Notes

1 The risks associated with nuclear power are both real and perceived, and they include both health and environmental risks. The perceived risks associated with nuclear power are generally higher than the objective risks assessed during the fifty years that nuclear power plants have been operating in the United States. These risks are discussed further later in the chapter.

2 The experiment was known as the Chicago Pile-1. The primitive reactor consisted of uranium pellets for the fuel, graphite blocks as the moderator, and rods covered with cadmium as the control rods. The reaction was supervised by Enrico Fermi and Leó Szilárd (U.S. DOE 2013).

3 Rural siting and isolation from populations have to be balanced with the location of the higher demand for electricity in urban settings and the corresponding need for distribution networks.

4 The Becquerel (Bq) is a measure of the radioactivity of a substance or item, rather than of the dose received, which is typically measured in rems.

5 Though siverts (Sv) have officially replaced the rem as the unit of radiation exposure, the rem is still commonly used. The conversion rate is 1 Sv is equal to 100 rems, or 1 millisivert (mSv) is equal to 100 millirems (mrem). A rem is a measure of ionizing radiation resulting in damage to the human body that is equivalent to the damage received by one Roentgen of high voltage x-rays (*MIT Tech Talk* 1994).

6 The United States Army had developed an experimental Stationary Low Power Reactor Number 1, or SL1, reactor. The reactor experienced a steam explosion and meltdown when the central control rod was improperly withdrawn. The

accident resulted in the abandonment of this type of reactor and a redesign of reactors so that the removal of a single control rod would not result in an out-of-control reaction (Tucker 2009).

7 Estimates of the precise number of deaths and post-radiation diseases vary. Thirty deaths and over a hundred radiation injuries are directly attributable to the accident. By 2005, over six thousand cases of thyroid cancer had been reported among residents of Belarus who were evacuated and permanently relocated. The average radiation exposure was estimated to be 120 mSv for recovery workers, 30 mSv for evacuees, and 9 mSv for people who have continued to live in the contaminated areas for the twenty years since the accident (UNSCEAR 2008).

8 The freeze in nuclear power is often attributed to the accident at the Pennsylvania Three Mile Island facility in March 1979. However, economic concerns had already stopped the expansion and adoption of nuclear power in the early 1970s.

9 The Vogtle power plant in Georgia is a two-unit pressurized water reactor (PWR) power plant with a combined operating capacity of 2400 MW. The units were built and commissioned in the 1980s. The plant originally had a forecast capital investment of $660 million, but it actually turned out to cost $8.87 billion (U.S. Energy Information Administration 2005; Gertner 2006). Two additional units are under construction at the plant. They received their construction and operating licenses in February 2012 and they are scheduled to enter service in 2017 and 2018 (Southern Company 2013).

10 Though utilities are constrained by what they are allowed to charge consumers for electricity, the underlying cost of electricity production is an important factor in electricity prices. Of course, the costs and benefits from nuclear power are dependent on the mix of energy sources within the jurisdiction. If a utility has only a relatively small amount of nuclear capacity, the effects will not be as significant.

References

Amundson, Michael A. 2004. *Yellowcake Towns: Uranium Mining Communities in the American West*, Boulder, CO: University Press of Colorado.

Anderson, Dennis and Matthew Leach. 2004. "Harvesting and Redistributing Renewable Energy: on the Role of Gas and Electricity Grids to Overcome Intermittency Through the Generation and Storage of Hydrogen," *Energy Policy*, 32, 14: 1603–1614.

Anderson, Kevin and Alice Bows. 2010. Reframing the Climate Change Challenge in Light of Post-2000 Emission Trends. In B. Launder and J. M. T. Thompson (eds), *Geo-Engineering Climate Change: Environmental Necessity or Pandora's Box?* (27–49). Cambridge, UK: Cambridge University Press.

ANS. 2012. *Fukushima Daiichi: ANS Committee Report*. LaGrange Park, IL: American Nuclear Society.

ASME. 2012. *Forging a New Nuclear Safety Construct: The ASME Presidential Task Force on Response to Japan Nuclear Power Plant Events*, New York, NY: The American Society of Mechanical Engineers.

Berube, Alan, Elizabeth Deakin, and Steven Raphael. 2006. Socioeconomic Differences in Household Automobile Ownership Rates: Implications for Evacuation Policy. Available at: http://gsppi.berkeley.edu/faculty/sraphael/berubedeakenraphael.pdf (accessed 24 June 2013).

Bezdek, Roger H. and Robert M. Wendling. 2006. "The Impacts of Nuclear Facilities on Property Values and Other Factors in the Surrounding Communities," *International Journal of Nuclear Governance, Economy, and Ecology,* 1, 1: 122–144.

Campbell, John L. 1988. *Collapse of an Industry: Nuclear Power and the Contradictions of U.S. Policy*, Ithaca, NY: Cornell University Press.

Chandel, Munish K., Lincoln F. Pratson, and Robert B. Jackson. 2011. "The Potential Impacts of Climate-Change Policy on Freshwater Use in Thermoelectic Power Generation." *Energy Policy*, 39, 10: 6234–6242.

Combs, Barbara and Paul Slovic. 1979. "Newspaper Coverage of Causes of Death," *Journalism & Mass Communication Quarterly,* 56, 4: 837–843, 849.

Covello, Vincent T. 1983. "The Perception of Technological Risks: A Literature Review," *Technological Forecasting and Social Change,* 23, 4: 285–297.

Dedman, Bill. 2011. Nuclear Neighbors: Population Rises Near US Reactors: Map of Census Data Shows a 17 Percent Increase in Residents within 10 miles in a decade, NBC News, 4/14/2011. Available at: www.nbcnews.com/id/42555888/ns/us_news-life/#.UkjPeBCK3–u (accessed 20 June 2013).

Dell, R.M. and D.A.J. Rand. 2001. "Energy Storage — a Key Technology for Global Energy Sustainability," *Journal of Power Sources,* 100, 1–2: 2–17.

Deutch, John M., Charles W. Forsberg, Andrew C. Kadak, Mujid S. Kazimi, Ernest J. Moniz, John E. Parsons, et al. 2009. *Update of the MIT 2003 Future of Nuclear Power*. Cambridge, MA: MIT Energy Institute.

Donn, Jeff. 2011. AP IMPACT: Populations Around US Nuke Plants Soar. Available at: www.ap.org/company/awards/part-iii-aging-nukes (accessed 20 June 2013).

Feeley III, Thomas J., Timothy J. Skone, Gary J. Stiegel Jr, Andrea McNemar, Michael Nemeth, Brian Schimmoller, et al. 2008. "Water: A Critical Resource in the Thermoelectric Power Industry," *Energy,* 33, 1: 1–11.

Freudenburg, W.R. 1988. "Perceived Risk, Real Risk: Social Science and the Art of Probabilistic Risk Assessment," *Science,* 242, 4875: 44–49.

Fthenakis, Vasilis and Hyung Chul Kim. 2010. "Life-cycle uses of water in U.S. electricity generation," *Renewable and Sustainable Energy Reviews,* 14, 7: 2039–2048.

Gamble, Hays B. and Roger H. Downing. 1982. "Effects of Nuclear Power Plants on Residential Property Values," *Journal of Regional Science,* 22, 4: 457–478.

Gardner, Gerald T. and Leroy C. Gould. 1989. "Public Perceptions of the Risks and Benefits of Technology," *Risk Analysis,* 9, 2: 225–242.

Gertner, Jon. 2006. Atomic Balm? *The New York Times*. Available at: www.nytimes.com/2006/07/16/magazine/16nuclear.html?_r=0 (accessed 7 April 2014).

Hecht, Gabrielle. 1998. *The Radiance of France: Nuclear Power and National Identity After World War II*, Cambridge, MA: The MIT Press.

Institute for Energy Research. 2012. Abundant Natural Gas Means Low Prices, Increased Trade Potential, *Institute for Energy Research*. Available at: www.instituteforenergyresearch.org/2012/04/19/abundant-natural-gas-means-low-prices-trade-potential/ (accessed 25 October 2013).

Jackson, Thomas. 2001. "The Effects of Environmental Contamination on Real Estate: A Literature Review," *Journal of Real Estate Literature,* 9, 2: 91–116.

Kamalick, Joe. 2013. Market outlook: US Poised for Century of Gas Abundance. Available at: www.icis.com/Articles/2013/04/26/9662821/market-outlook-us-poised-for-century-of-gas-abundance.html (accessed 25 October 2013).

Kiel, Katherine A. and Katherine T. McClain. 1995. "House Prices during Siting Decision Stages: The Case of an Incinerator from Rumor through Operation," *Journal of Environmental Economics and Management,* 28, 2: 241–255.

Kinnaman, Thomas C. 2011. "The Economic Impact of Shale Gas Extraction: A Review of Existing Studies," *Ecological Economics,* 70, 7: 1243–1249.

Kirkwood, Craig W. 1982. "A Case History of Nuclear Power Plant Site Selection," *The Journal of the Operational Research Society,* 33, 4: 353–363.

Klein, Dale and Michael Corradini. 2012. *Fukushima Daiichi: An ANS Committee Report*, LaGrange Park, IL: American Nuclear Society.

Lazonick, William. 2008. "The Quest for Shareholder Value: Stock Repurchases in the US Economy," *Louvain Economic Review,* 74, 4: 479–540.

Mahaffey, James. 2009. *Atomic Awakening: A New Look at the History and Future of Nuclear Power*, New York, NY: Pegasus Books LLC.

Malin, Stephanie A. and Peggy Petrzelka. 2010. "Left in the Dust: Uranium's Legacy and Victims of Mill Tailings Exposure in Monticello, Utah," *Society and Natural Resources,* 23, 12: 1187–1200.

McCaffrey, David. 1990. *The Politics of Nuclear Power: A History of the Shoreham Nuclear Power Plant*, Boston, MA: Kluwer Academic Publishers.

—— 2004. *The Politics of Nuclear Power: A History of the Shoreham Nuclear Power Plant*, New York, NY: Springer Science Business Media BV.

Medvedev, Grigori. 1991. *The Truth about Chernobyl*, New York, NY: Basic Books, Inc.

Medvedev, Zhores A. 1990. *The Legacy of Chernobyl*, New York, NY: W.W. Norton and Company.

MIT Tech Talk. 1994. "Radiation, How Much is Considered Safe for Humans?," *MIT Tech Talk,* 38, 18: January 5, 1994.

Nuttall, William J. 2004. *Nuclear Renaissance: Technologies and Policies for the Future of Nuclear Power*, New York, NY: Taylor and Francis.

Riddel, Mary, Christine Dwyer, and W. Douglass Shaw. 2003. "Environmental Risk and Uncertainty: Insights from Yucca Mountain," *Journal of Regional Science,* 43, 3: 435–458.

Ringholz, Raye. 2002. *Uranium Frenzy: Saga of the Nuclear West*, Logan, UT: Utah State University Press.

Slovic, Paul, Baruch Fischhoff, and Sarah Lichtenstein. 1980. Facts and Fears: Understanding Perceived Risk. In R. Schwing and J. Albers (eds), *Societal Risk Assessment: How Safe is Safe Enough?* (181–216). New York, NY: Penum.

Society, Health Physics. 2013. Regulatory Dose Limits. Available at: http://hps.org/publicinformation/ate/faqs/regdoselimits.html (accessed 20 September 2013).

Southern Company. 2013. Plant Vogtle Units 3 and 4. Available at: www.southerncompany.com/what-doing/energy-innovation/nuclear-energy/pdfs/vogtle-nuclear-brochure.pdf (accessed 29 June 2013).

Sovacool, Benjamin K. and Kelly E. Sovacool. 2009. "Identifying Future Electricity-water Tradeoffs in the United States," *Energy Policy,* 37, 7: 2763–2773.

The Economist. 2013. Difference Engine: Fuel for the Future?, *The Economist.* Available at: www.economist.com/blogs/babbage/2013/05/natural-gas (accessed 25 October 2013).

Tucker, Todd. 2009. *Atomic America: How a Deadly Explosion and a Feared Admiral Changed the Course of Nuclear History*, New York, NY: Free Press.

U.S. Census. 2013. USA Counties. Available at: http://censtats.census.gov/usa/usa.shtml (accessed 28 June 2013).

U.S. DOE. 2013. The Manhattan Project: An Interactive History. Available at: http://web.archive.org/web/20101122183641/ www.cfo.doe.gov/me70/manhattan/cp-1_critical.htm (accessed 28 June 2013).

U.S. Energy Information Administration. 2005. U.S. Nuclear Power Plants – Vogtle. Available at: www.eia.gov/404r.cfm?v=http://www.eia.gov/cneaf/nuclear/page/at_a_glance/reactors/vogtle.html (accessed 28 June 2013).

—— 2012. *Annual Energy Outlook 2012 with Projections to 2035, June 2012, DOE/EIA-0383(2012)*. Washington, DC: Energy Information Administration.

U.S. Nuclear Regulatory Commission. 2011. *Occupational Radiation Exposure at Commercial Nuclear Power Reactors and Other Facilities 2011, Forty-Fourth Annual Report, NUREG-0713 Volume 33*. Washington DC: Nuclear Regulatory Commission.

UNSCEAR. 2008. The Chernobyl accident: UNSCEAR's assessments of the radiation effects. Available at: www.unscear.org/unscear/en/chernobyl.html (accessed 20 September 2013).

Walker, J. Samuel. 2004. *Three Mile Island: A Nuclear Crisis in Historical Perspective*, Berkeley and Los Angeles, CA: The University of California Press.

—— 2009. *The Road to Yucca Mountain: The Development of Radioactive Waste Policy in the United States*, Berkeley, CA: University of California Press.

Watts, Richard. 2012. *Public Meltdown: The Story of the Vermont Yankee Nuclear Power Plant*, Amherst, MA: White River Press.

Weber, Jeremy G. 2012. "The Effects of a Natural Gas Boom on Employment and Income in Colorado, Texas, and Wyoming," *Energy Economics*, 34, 5: 1580–1588.

World Nuclear Association. 2011. The Nuclear Renaissance. Available at: www.world-nuclear.org/info/Current-and-Future-Generation/The-Nuclear-Renaissance/#.UnRHGxCK3–s (accessed 26 September 2013).

—— 2013a. The Economics of Nuclear Power. Available at: www.world-nuclear.org/info/Economic-Aspects/Economics-of-Nuclear-Power/#.UlQ4XhCK3–s (accessed 6 October 2013).

—— 2013b. Nuclear Radiation and Health Effects. Available at: www.world-nuclear.org/info/Safety-and-Security/Radiation-and-Health/Nuclear-Radiation-and-Health-Effects/#.UcRsjZyPb7I (accessed 29 June 2013).

World Nuclear News. 2009. Obama Dumps Yucca Mountain. Available at: www.world-nuclear-news.org/newsarticle.aspx?id=24743 (accessed 6 October 2013).

6 Socioeconomic Issues and Biofuel Energy

Harry de Gorter and Dusan Drabik

Introduction

Biofuel policies have arisen as a confluence of measures which are a subset of energy, environmental and agricultural legislation designed to obtain multiple goals, including energy security, environmental quality (reduced local air pollution and GHG emissions), improved farm income and rural development, and induced technological change. Other stated political goals include increased domestic energy production that creates 'green' jobs, and foreign exchange savings with reduced crude oil imports.

The effectiveness of biofuel policies in achieving these various policy goals has been questioned (for surveys, see de Gorter and Just 2010a; de Gorter et al. 2013d; FAO 2013). A series of studies have questioned the savings in GHG emissions by analyzing the indirect land use change due to higher crop prices (e.g. Searchinger et al. 2008; Khanna et al. 2011) while others have questioned the benefits of reduced GHGs due to indirect output use change in the fuel market (e.g. Drabik and de Gorter 2013).

Although biofuel production and consumption are mostly concentrated in the United States, Europe and Brazil, more than sixty countries have implemented biofuel policies (Agra Europe 2013). The most important policies are blending mandates (e.g. in the United States) or mandatory targets (e.g. in the European Union), subsidies aimed at supporting biofuel consumption (i.e. tax credits or tax exemptions) and production (e.g. ethanol production), subsidies on biofuel feedstock production (e.g. U.S. corn), and binary sustainability standards with indirect land use measures that determine whether or not a biofuel can be counted toward a mandate (Rajcaniova et al. forthcoming). Although the economics of each of these individual policies are unique, they all result in a lower crude oil price and higher feedstock prices (de Gorter and Just 2009a, b; Stoft 2010; Cui et al. 2011; Drabik 2011; Rajagopal et al. 2011; Lapan and Moschini 2012).

The effect of biofuel policies is to lower average fuel prices to consumers except with a mandate alone, when prices could go up or down (de Gorter and Just 2009b; Drabik 2011; Lapan and Moschini 2012; Rajagopal and Plevin 2013). The intuition for this is the fact that a mandate acts as a tax on the gasoline market, thereby always causing gasoline market prices to decrease. Gasoline consumers pay a higher price for gasoline (to accommodate the high ethanol price), and gasoline producers obtain a lower price. So the mandate is acting both as a monopolist against gasoline consumers and as a monopsonist against gasoline producers. It is possible that the revenues extracted from gasoline producers are so high (when the gasoline supply curve is sufficiently inelastic) that total fuel production (and hence consumption) increases, implying that the fuel price decreases. Consumers still pay a higher price for gasoline, but they pay a lower fuel price once ethanol is included.

Separately, biofuel policy is a double-edged sword in achieving improved farm income/rural development even as the tax on value-added agriculture (e.g. livestock, dairy and poultry) reduces the incomes of farmers and may also reduce economic growth in rural areas net of the economic growth due to biofuel production. Furthermore, the new high price era due to biofuels itself induces politicians to increase support prices and create new policy instruments to transfer taxpayer dollars to farmers, so that current Farm Bill deliberations have changed dramatically.

Biofuels policy has had such a big impact on food commodity prices because it created a direct link between crop and energy prices for the first time ever; before the policy was introduced, energy prices only affected crop prices through input costs. The biofuel-crop price multiplier is very high. For example, for every 1 cent per gallon increase in the ethanol price, corn prices increase by about 4 cents per bushel (de Gorter and Just 2008). Because crops compete for the same land and are substitutable in demand, all grain/oilseed prices went up together,[1] and all other biofuel production worldwide therefore experienced additive effects. The near doubling of corn prices over the five months beginning in October 2006 was the catalyst, a delayed response to a doubling in one year of an already escalating U.S. production capacity for ethanol. The large corn-ethanol price multiplier, coupled with comparable price links in biodiesel and the Brazilian sugarcane-ethanol markets, has been the main cause of the turmoil in world grain/oilseed prices witnessed in the past half-decade.

There are many policy questions surrounding biofuels, with their impact on food commodity prices and GHG emission reductions being the most prominent. Hence, we focus on these two broad issues in this chapter. The remainder of this chapter is organized as follows. In the next section, we analyze the impact of biofuel policies on food grain/oilseed prices and evaluate the new counterfactual: crop prices are now tied to biofuel policies. The third section analyzes the expected relationships between food grain/oilseed, biofuel (ethanol and biodiesel) and energy (gasoline/diesel and crude oil) prices. Next we discuss the welfare effects on the environment, and the final section concludes by looking at how biofuels affect rural development and the politics of the new Farm Bill.

Impact of Biofuel Policies on Food Commodity Prices and Price Volatility

In this section, we discuss the importance of biofuel policies in explaining the recent rise in food commodity prices. In particular, biofuel policies caused a direct link between crop (cereals and oilseeds) and biofuel (ethanol and biodiesel) prices for the first time ever. Figure 6.1 indicates that a close link between world corn and U.S. ethanol prices did not exist before September 2006. However, the effects of the *de facto* ban on MTBE use, a petroleum-based competitor to ethanol, and high oil prices after 2005, which activated the dormant blender's tax credit (in place since 1978), caused corn and ethanol prices to become closely linked. These events were the key environmental and energy policies that precipitated the food grain price links to ethanol and the ensuing increase in crop price levels and volatility. As it will become clear shortly, the link between corn and ethanol (and oilseeds and biodiesel) is necessary but not sufficient to explain the past development of corn prices. We first have to explain how the ethanol market price is determined.

Historically, the U.S. ethanol market price has been determined either by the blender's tax credit or the blend mandate. (We say that a policy is binding if it determines the market price of a biofuel.) These two policies have unique economics which underlie the behavior of the food grain and oilseed prices. If the tax credit is binding, the ethanol market price is given by:

$$P_E = \lambda P_G - (1-\lambda)t + t_c \tag{1}$$

where $\lambda = 0.7$ and denotes miles traveled per gallon of ethanol relative to gasoline, P_G denotes the market price of gasoline, t denotes the fuel tax, and t_c represents the blender's tax credit. Note that the same equation holds for the case when the tax credit is not present (i.e. $t_c = 0$). An important implication of equation (1) is that if the tax credit is binding, then the ethanol price (and hence also the corn price) is directly linked to the gasoline (and therefore also to the crude oil) price.

If the mandate is binding, then the ethanol market price must be higher than the price determined by equation (1). The price formation for that case is derived in de Gorter and Just (2009b). The key difference between how the ethanol price is determined under the binding blend mandate and under the tax credit is that the ethanol price is not directly linked to the gasoline price, and in fact it can float up and away from it.

The two channels through which the ethanol price is established lead to two states of nature for ethanol and corn prices. In the first state, the two prices (locked to each other) are the lowest when ethanol prices are locked onto gasoline (oil) prices (in which case corn and crude oil prices are strongly positively correlated). In the second state (which occurs when the

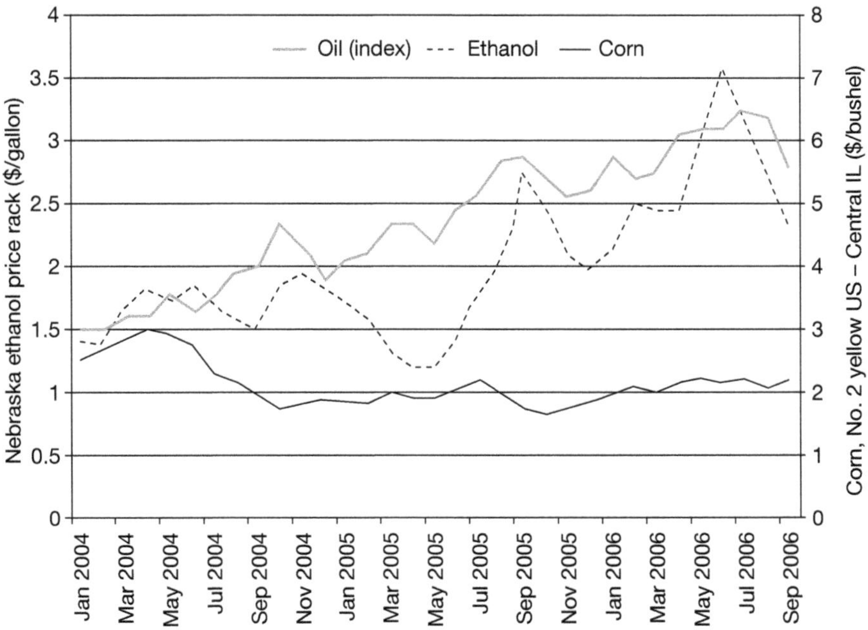

Figure 6.1 Oil, Ethanol and Corn Prices, Jan 2004 to Sept 2006.
Source: Rausser and de Gorter (2013).

"mandates"[2] become binding), ethanol, and hence corn prices, float up and away from gasoline (oil) prices. In the latter state, ethanol (corn) prices move higher when they become delinked from gasoline prices. In that case, corn prices are weakly negatively related to crude oil prices *ceteris paribus*.[3]

Why are the two states of nature important? If corn prices are locked onto crude oil prices through the tax credit, then supply/demand shocks in corn markets have no impact on corn prices (unless *a change* in ethanol production could affect world crude oil prices, which is not very likely). Factors like input costs or inventory levels in this case do not matter. Only when biofuel prices become delinked from energy prices can crop supply/demand shocks affect corn prices.

Although corn prices did not move for thirty-three months leading up to October 2006, ethanol and oil prices doubled. Then corn-ethanol prices became linked with the U.S. farm price of corn, increasing by 88 percent (KC #2 white corn 107 percent) in five months, precipitating the Mexico Tortilla Crisis in January 2007, the India wheat import ban one month later (the first of many developing country policy responses) and the doubling of the U.S. mandate in December 2007 (see Figure 6.2).[4] By then, corn and ethanol prices were in lockstep and they continue so to this day, as crop-biofuel price links also developed between soybean/canola oil and biodiesel, and between sugarcane-ethanol and sugar in Brazil. This ushered in a new era of high and volatile food grain/oilseed prices that has existed to this day.

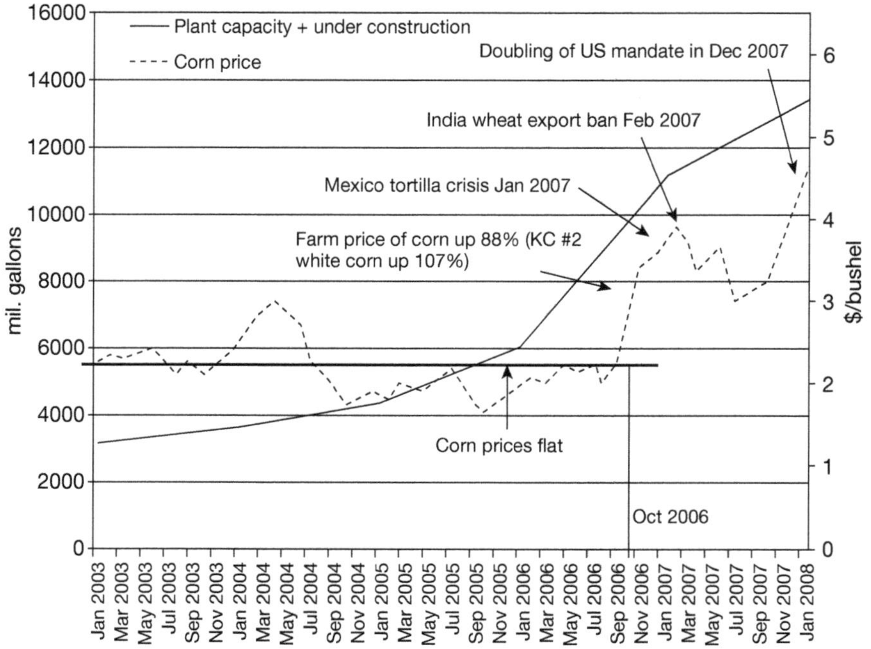

Figure 6.2 Ethanol Plant Capacity and U.S. Farm Price of Corn.
Source: adapted from Rausser and de Gorter (2013).

Now that we know how the ethanol price can be determined, we are in a position to investigate how sensitive the corn price is to changes in the ethanol price. To that end, we present an algebraic representation of the corn-ethanol price referred to above (see e.g. Drabik 2011 for the derivation).

$$P_C = \frac{\beta}{1 - r\gamma}\left(P_E - c_0\right) \tag{2}$$

The corn-ethanol price link is determined by several key factors: how much ethanol is produced from a bushel of yellow corn (β); how much co-product – called dried distillers grains with solubles, DDGS – is produced in the process of producing ethanol (γ) (DDGS is subsequently used as a corn substitute to feed animals); the relative prices of DDGS and corn (r); and the cost associated with processing corn to ethanol. At this point it is important to realize that the corn-ethanol price multiplier (the fraction term on the right-hand side) is very high and means empirically that for every 1 cent per gallon increase in ethanol prices, corn prices increase about 4 cents per bushel.[5]

How well does equation (2) forecast corn prices? The evidence is given in Figure 6.3, where there are three early time periods of overly high prediction due to ethanol production capacity constraints: the ramp up of ethanol production in 2007, the restructuring of bankrupt ethanol plants in 2009–10, and the tax credit subsidizing exports in 2011. The drought in 2012 caused the reverse; we under-predicted corn prices.

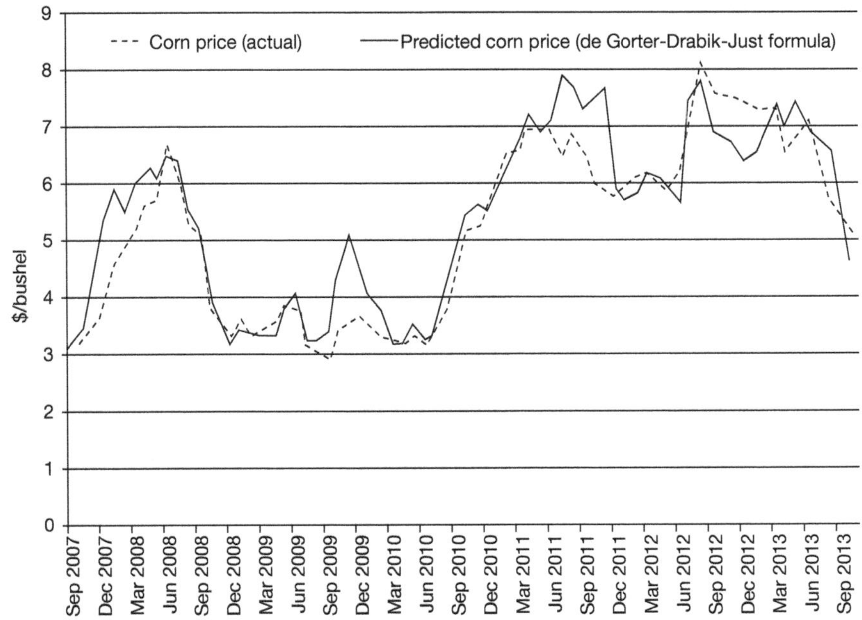

Figure 6.3 Predicted versus Actual Corn Prices.
Source: derived from Drabik (2011).

Most economic studies of biofuel policies model their impact as a shift in crop demand and ignore the price link between crops and biofuels (equation [2]). These studies analyze the food grain price increases through the lens of traditional economic analyses such as the effects of exchange rates, macroeconomic policies and shocks, speculation, commodity supply/demand trends and shocks, the behavior of stockholders, and the like.[6] The overall conclusion of that strand of literature is that biofuels had an impact on food grain commodity prices, but they were only one of many factors in a "perfect storm". Moreover, that literature has also concluded that biofuel policy is only partly responsible for the fraction of food grain commodity price increases that is due to biofuels.

Unlike the literature above, our conceptual approach, presented below, is to carefully analyze the specific relations between the oil, biofuel and feedstock prices domestically and internationally, depending on which biofuel policy is binding. We recognize that crop and biofuel prices are directly linked through some very specific structural relationships, and we account for the conditions that create or break these relationships. Although these price links also exist for other crops and biofuels, and in other countries, biofuel prices between countries are often delinked due to barriers to trade. However, empirical observations show that relative prices of food grains and oilseeds have maintained their historical relationship due to competition for land and substitution in demand. We call this stable relationship the *law of one relative international price* for food grains and oilseeds.[7]

To answer the key question of this chapter, namely how much of the food grain/oilseed price increase is due to biofuel policies given the tight crop-biofuel price links with high multipliers and the two states of nature, we need three additional concepts:

1. The ethanol *price premium* over the free market ethanol price[8] (i.e. the observed ethanol price less the free market price) is very high because one gallon of ethanol delivers only 70 percent of the miles traveled compared to gasoline. Furthermore, 30 percent of the volumetric fuel tax has to be passed on to consumers by blenders, thereby lowering the free market ethanol price (we call this the volumetric fuel tax "penalty" on ethanol). We estimate this price premium to average around 60¢/gal (or approximately $2.40/bu if measured in prices of corn) for the 2007–2011 crop years.
2. It is important to understand that historically, the intercept of the ethanol supply curve has been above the free market ethanol price. We call this situation '*water*' in ethanol price premium.[9] Intuitively, the minimum price that the producer requires to start ethanol production is higher than what the consumer is willing to pay given that conventional gasoline is available and yields more miles per gallon than ethanol.
3. The *new counterfactual* for ethanol (and hence corn) prices is not the world without biofuels mandates, but rather the world with tax credits, in which case ethanol (corn) prices are locked onto gasoline (crude oil) prices as per equation (1).[10] This means the lowest ethanol and hence corn prices can go is very high compared to historical prices. Non-biofuel factors that shift the corn supply and demand curves have little impact with this counterfactual unless crude oil prices are affected. Hence, the only way non-biofuel supply/demand factors can affect corn prices is if they activate dormant mandates such that the ensuing ethanol market price is higher than under the tax credit. Empirically, we find that such a "mandate price premium over the tax credit" averaged 60¢/bu over the 2007–2011 time period.

Now we are in a position to present empirical results on the net impact of biofuel policies on crop prices compared to the new counterfactual. For ease of exposition, we use the following notation:

P_C: observed corn price;

$P_C^{t_c}$: the predicted corn price assuming the tax credit was binding throughout the time period using equations (1) and (2);

P_C^{lowest}: the lowest corn prices can go where P_{NE} represents a floor[11] and the predicted corn price with the tax credit is not allowed to be higher than the actual corn price.

Therefore:

$$P_C^{lowest} = P_{NE} \text{ if } P_C^{t_c} < P_{NE}; = P_C \text{ if } P_C^{t_c} > P_C; \text{ otherwise} = P_C^{t_c}$$

The corn price premium over the tax credit is defined as the difference between the actual corn price and the lowest corn prices can go (i.e. $P_C - P_C^{lowest}$).[12]

Table 6.1 gives the results. The average corn price for 2000/01–2005/06 is \$2.10/bu and for the period 2005/06–2011/12 it is \$4.68/bu. The increase is thus \$2.58/bu or 123 percent. The lowest corn prices could go in the period 2005/06–2011/12 averages \$4.10/bu, leaving an average mandate premium over the tax credit to be \$0.58/bu (=\$4.58–\$4.10), or 23 percent of the corn price increase.

When the tax credit is binding, supply/demand shifts cannot significantly affect the corn price. Therefore, 100 percent of the corn price change is due to the tax credit if it is binding all of the time. However, it was not, as the data in Table 6.1 shows. Because the average mandate price premium over

Table 6.1 How Much of the Corn Price Increase is due to Non-biofuel Factors?[a]

	Prices	*Value \$/bushel*	
		Profits	No Profits
1	Baseline corn price (2000/01 to 2005/06)	2.10	
2	Average corn price (2006/07 to 2011/12)	4.68	
3	Increase in corn price	2.58 (123 percent)	
4	The lowest corn prices can go[b]	4.10	4.22
5	Premium over the tax credit[c]		
6	Gross	0.58 (23 percent of price increase)	0.46 (18 percent of price increase)
7	Deduct when the mandate is binding even w/o S/D shifts[d]	–0.03 (21 percent of price increase)	–0.04 (16 percent of price increase)
8	How much of ethanol price premium over the tax credit is due to non-U.S. corn-ethanol biofuels[e]	To be determined	
9	Net impact of non-biofuel factors on corn prices	To be determined	

Source: derived from Drabik (2011).

[a] Analysis looks at U.S. corn-ethanol market only and so requires the "law of one international relative price" to hold in order to extend to all food grain/oilseed price changes.

[b] Average predicted corn price defined by Equation (2), adjusted to disallow for observations: (a) exceeding actual corn prices; and (b) falling below the non-ethanol corn price P_{NE}.

[c] The premium over the tax credit is defined as the observed corn price minus the lowest corn prices can go (note the tax credit expired at the end of 2011).

[d] Accounts for the amount by which the premium over the tax credit is higher than the gap between the baseline corn price and the lowest corn prices can go.

[e] For example, U.S. biodiesel takes land out of corn production and ethanol from wheat increases demand for feed corn. Other biofuel policies are therefore mistaken as non-biofuel supply/demand shifts and so an adjustment has to be made. Note that non-U.S. corn-ethanol production in OECD countries is half as much but uses more land because U.S. corn-ethanol is by far the most efficient biofuel (excluding Brazilian sugarcane-ethanol).

the tax credit represents 23 percent of the corn price increase, the rest (77 percent) must be due to the tax credit alone (unless some ethanol would have been produced under free markets, but the link between corn and ethanol prices would never have occurred without energy and environmental policies, so we assume that all ethanol production is due to ethanol policy).

So what are the implications of the above for Mitchell's (2008) widely cited paper, in which he concluded: "…most of the remaining 70–75 percent increase in food commodities prices was due to biofuels and the related consequences of low grain stocks, large land use shifts, speculative activity and export bans"? We note that in cases where the corn prices were lowest when they were locked onto energy prices, the statement "…the related consequences of low grain stocks, large land use shifts, speculative activity and export bans" is probably immaterial and can only impact the mandate premium.

We have already explained that with only the price premium due to mandates over the tax credit, supply/demand shocks in the corn market can affect the ethanol and corn prices.[13] This premium is estimated to be 23 percent of the observed increase in corn prices. However, before attributing this premium to corn supply/demand shifts, we first need to make two other adjustments which can best be illustrated by means of Figure 6.4. The heavy black line denotes the observed corn price; the dashed line represents the mandate premium over the tax credit (i.e. the observed ethanol price less the price predicted by equation [1]); the dotted line denotes the baseline (average) corn price of $2.10/bu; and the tinted line denotes the lowest corn price possible, P_C^{lowest}, as defined above.

The first adjustment relates to a situation when the mandate is binding even without supply/demand shifts. This occurs when the amount by which the premium over the tax credit is higher than the gap between the lowest corn prices can go – P_C^{lowest} in Figure 6.4 (where P_C^{lowest} is adjusted such that it cannot go below P_{NE} and $P_C^{t_c}$ cannot go above actual corn prices) – and the baseline corn price (average corn price of $2.10 before the advent of ethanol).

To calculate the first adjustment factor, we proceed in four steps. First, we calculate the difference between the minimum corn prices and the baseline price; denote this distance as *a*. Then we calculate the difference between the premium over the tax credit and distance *a*. Next, we sum over all the positive resultant differences. Finally, we divide the sum by the total number of observations in the sample. The obtained adjustment factor is $0.03/bu, meaning that the net share of the corn price increase due to the mandate premium over the tax credit falls to 21 percent (row 7 in Table 6.1).

To put this last figure into perspective and to highlight the importance of our results, let us consider the results of Roberts and Schlenker (2013) who, like most papers analyzing the impact of biofuels on grain and oilseed markets, do not model the biofuel policy or the crop-biofuel price links. These authors conclude that grain and oilseed prices increased by 30 percent

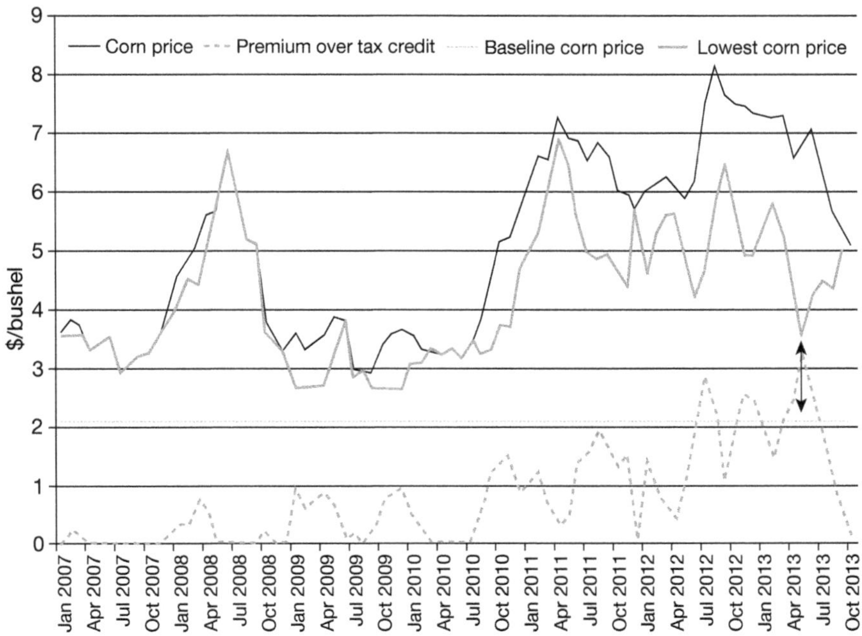

Figure 6.4 Upper Bound of non-U.S. Ethanol Effects on Corn Prices.
Source: derived from Drabik (2011).

in the 2010/11 crop year due to corn-ethanol. If one adjusts for the fact that 24 percent[14] of the corn processed into ethanol was recycled as (equivalent) corn, then 22.7 percent of the grain and oilseed price increase is due to biofuels, leaving 77.3 percent to supply and demand shifts. However, we show above that by recognizing the new reality of the corn-ethanol-gasoline price links, the effect of supply and demand shifts on corn prices in 2010/11 was 21 percent at most. The change in consumer surplus due to biofuels in Roberts and Schlenker's (2013) results, adjusted for our more accurate recycling estimate, is a loss of $136.2 billion. However, the loss increases to $410 billion (more than three times the loss estimated by Roberts and Schlenker [2013])[15] if one considers the new corn price counterfactual.

The second (conceptual) adjustment relates to the effects of the shifts in corn supply and demand curves due to biofuels production other than U.S. corn-ethanol (i.e. U.S. biodiesel and biofuel production in the rest of the world). Non U.S. corn-ethanol biofuel production in OECD countries is half of the U.S. level, and on proportionately more land. This means more land devoted to corn (and soybeans) and less to other crops, shifting the U.S. corn supply curve up, while wheat used for ethanol production in the United Kingdom, for example, increases the demand for corn feed. Therefore, the net impact of non-biofuels supply/demand shifts in corn and other crops may be zero!

In summary, we have shown that U.S. corn-ethanol policies explain 79 percent of the corn price increase from 2007 to 2011, given the new counterfactual in which corn prices are tied to the gasoline (crude oil) price through the tax credit and the ethanol-gasoline price link. Since the law of one relative international price of food grains and oilseeds holds (because of competition for land and substitution in demand), 79 percent of all food grain and oilseed price increases are due to biofuel policies. However, OECD biofuel production excluding U.S. corn-ethanol is half as much again, and because U.S. biodiesel production shifts the corn supply curve up or wheat used for ethanol in Canada shifts the feed demand curve for corn out, much of this premium may be due to biofuels production outside of U.S. corn-ethanol. Hence, we come to our conclusion that biofuels policies explain most of the food grain and oilseed price increases. The price links among crops and between crops and biofuels, their magnitude and the new counterfactual that crop prices can go no lower than that when tied to oil prices, are very real, with important policy implications.

Crop-Biofuel-Energy Price Links: How to Determine the Role of Biofuel Policies?

In this section, we focus on the price links between grain/oilseed-biofuel (ethanol and biodiesel) and energy (gasoline/diesel and crude oil). In particular, we explain that these individual price relationships are dependent on the policy regime (i.e. tax credit or a mandate binding). Moreover, changes in the relationships can be affected by changes in regimes and policies as well as by prevailing market conditions. We will account for two types of capacity constraints—ethanol production capacity constraints and ethanol blending capacity constraints—which have caused ethanol prices to fall below gasoline prices on a miles per gallon basis for three short time periods in the past.

A burgeoning literature analyzes the dynamic linkages between crop, biofuel and energy prices using statistical time-series models to determine the influence of higher energy prices on crop prices (see Zilberman et al. 2013 and Serra and Zilberman 2013 for surveys).[16] This literature relies on statistical models used in the finance literature on volatility, such as vector-autoregression estimation and the use of Granger tests for causality. However, in our view this is only an indirect test of the impact of biofuels on crop prices, and it provides little economic insight as to how specific policies affect prices because these methods have so far been unable to control for the switch in regimes between a tax credit/tax exemption and a mandate (see our discussion of Figure 6.5).

However, biofuel prices did affect crop prices through the price linkages (see equation [2]). Furthermore, this econometrics literature is inconclusive about the role of biofuel policies because it fails to recognize that the crop-biofuel link began in October 2006 (the time-series analysis in this literature typically ends in 2007 or 2008). This literature has also not controlled for the switch in regimes between a tax credit/tax exemption and a mandate, or when U.S. biofuel prices are determined on world markets through exports.

Using Figure 6.5, we summarize the various crop-biofuel-energy price relationships that should be taken into account in an econometric analysis. We use U.S. corn-ethanol as an example and consider two policy regimes: the tax credit (corn prices are locked onto crude oil prices) and the blend mandate (corn prices float up and away). This distinction between the two policy regimes is very important because (*i*) the causality between corn and ethanol prices dependents on the policy regime, and (*ii*) the impact of supply/demand shocks on ethanol and hence corn prices also depends on the regime.

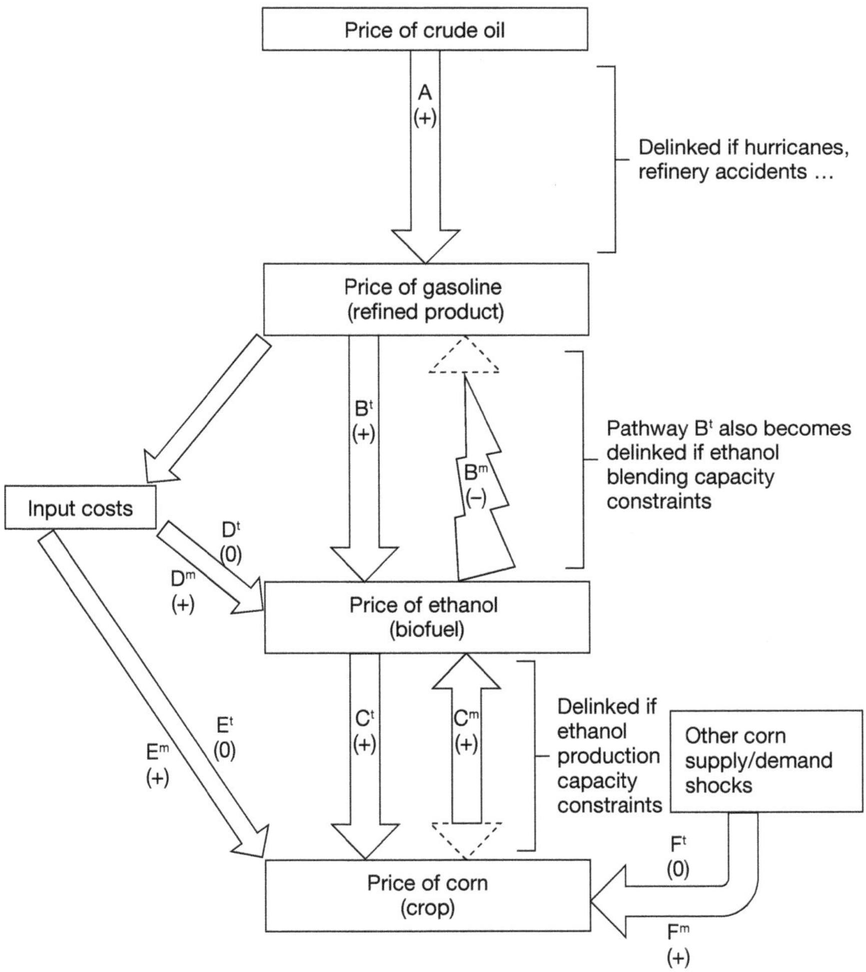

Figure 6.5 Energy-Biofuel-Crop Price Relationships.
Source: The Theory of Biofuels as developed by de Gorter, Drabik, and Just.

Note: The pathways with a superscript "t" are when a tax credit (exemption in the rest of the world) is binding. The superscript "m" denotes when a mandate is binding.

Oil-gasoline price link: Regardless of whether the mandate is binding or not, the relationship between the price of oil and the wholesale market price of gasoline remains the same: they are directly and positively related (Pathway A in Figure 6.5), with the coefficient depending on a number of technical parameters including seasonality. However, the price link can be broken, for example by hurricanes, thereby impacting gasoline and hence ethanol and corn price volatility.

One key aspect is the extent to which increased ethanol supplies have reduced gasoline and hence crude oil prices, and affected the "crack spread" (the margin for oil products) (Du and Hayes 2009). One should analyze the extent to which not only the crack spread has been affected (Knittel and Smith, forthcoming) but also diesel prices (Cui et al. 2011) and the international trade in petroleum products.

Oil-ethanol price link: If ethanol prices are locked onto gasoline and hence crude oil prices, an increase in oil price increases the price of gasoline, and therefore ethanol price increases as well, governed by equation (1), developing along Pathway B^t where the coefficient of price transmission depends on relative miles per gallon achieved of ethanol versus gasoline.[17] Gasoline price determines ethanol price (with a significantly smaller feedback effect—not shown in Figure 6.5—where increased ethanol production can lower gasoline and hence oil prices). An increase in oil price increases the costs of processing ethanol (Pathway D^t, where c_0 in equation [2] increases so corn prices decline through pathways C^t and C^m in Figure 6.5), but the price of ethanol is locked onto the gasoline price so ethanol prices do not change.

If the *blend mandate is binding* an increase in oil price increases the price of gasoline, so the fuel (blend of gasoline and ethanol) price increases, causing both fuel consumption and hence ethanol prices to decline (Pathway B^m).[18] This negative effect is not large when the blend ratio is around 10 percent. Ethanol blending constraints cannot occur with oil price declines in this situation. In this case, gasoline price has a negative effect on ethanol price. Equation (1) above does not hold: ethanol prices float up and away from gasoline prices.

An increase in oil price increases the costs of processing ethanol (Pathway D^m)—as c_0 in equation (2) increases, so the price of ethanol decreases relative to gasoline because a higher oil price means a higher gasoline price, and a higher gasoline price means a lower ethanol price under a binding mandate. This occurs independent of the corn-ethanol price link (unless one incorporates the feedback effect of ethanol production [prices] on gasoline/oil prices). However, one does have to analyze Pathways D^m and B^m simultaneously.

Ethanol-corn price link: Regardless of what policy is binding, the corn-ethanol price link (Pathways C^t and C^m) is governed by equation (2). The causality is strictly positive from ethanol to corn prices for Pathway C^t, but it is mostly from corn to ethanol in Pathway C^m in Figure 6.5. Gasoline-ethanol and ethanol-crop price link[19] can be broken by production and blending capacity constraints respectively (in which case corn prices are not as volatile or as high as they might otherwise have been).

Oil-corn price link: If the corn price is locked onto oil prices (tax credit binding in the past), an increase in the oil price has no direct impact on corn prices as increased costs of corn production simply shift the supply curve up and change the allocation of corn among uses, including a reduction of corn allocated to ethanol production along Pathway E^t.[20] However, oil prices have an indirect impact on corn prices: Pathway A to B^t to C^t has corn prices increasing, but Pathway D^t has c_0 increasing, causing corn prices to decline (the net effect on corn prices is most likely positive because higher oil prices affect ethanol production costs indirectly through natural gas and electricity prices etc.).

If the *mandate is binding,* the direct effect of higher oil (gasoline) prices on corn input costs, *ceteris paribus*, has corn and oil prices moving in opposite directions: Pathway C^m (see explanation for Pathway B^m).[21] If oil prices increase, corn input costs go up and so corn prices rise, taking ethanol prices with them (Pathway E^m).

Effects of other corn supply/demand shocks on the corn price: If the tax credit is binding, any shift in corn supply or demand will only change the allocation of corn to different uses, and corn prices are tied to gasoline (oil) prices: Pathway F^t (same reasoning as with Pathway E^t). If the *mandate is binding,* any shift in corn supply or demand will also affect corn prices, taking ethanol prices with them: Pathway F^m (same reasoning as with Pathway E^m).

The discussion so far highlights how the relationships between energy, biofuel and feedstock prices are nuanced, but nevertheless important in understanding the impact of biofuels (and agricultural policies) on price levels and volatility. With a binding mandate, there are four negative effects and one positive effect of a higher oil price on corn prices.[22] Therefore, simply analyzing the crude oil-corn price link misses the complexities of the crop-biofuel-energy price linkages—many of the mechanisms or pathways have to be estimated directly, conditional on regimes and whether there are any capacity constraints, as emphasized by Abbott (2012).

To emphasize the importance of the discussion above, consider a recent paper by Kristoufek et al. (2013) on the econometrics of the energy-biofuel-crop price relationships, which extends the literature by looking at non-linear relationships and concludes that the corn-ethanol (and biodiesel-rapeseed oil) price links did appear in 2007–08 but returned to pre-food crisis levels afterwards. Our analysis suggests the link remains very strong and any de-linking can be explained by our discussion of Figure 6.5 above.

Biofuels policies do not only affect crop price levels but also their volatility, since U.S. biofuels policies interact with each other and also with biofuels policies in the rest of the world, thereby increasing the complexity of the economics of commodity-price volatility. The interaction effects not only depend on the source of the shock (oil prices versus crop supply/demand shocks) or which biofuel policy determines the world biofuel market price (e.g. tax credit versus mandate), but also on the interactions across the various environmental, energy and agricultural policy instruments within a country as well as across countries. One example of this occurred in 2011 when the United States was the largest exporter of ethanol in the world, including Brazil, and so the two countries' ethanol prices were linked, causing market shocks in the sugar market to reverberate in corn markets. The tax credit with exports increased U.S. ethanol prices, and when the tax credit expired, U.S. and Brazilian ethanol prices both plummeted.[23]

Welfare Economics of Biofuel Policies

As both the mandate and the tax credit have frequently been used to stimulate biofuel production, it is useful to compare their welfare effects for the same quantity of ethanol consumption; de Gorter and Just (2010b) show that when the fuel tax is less than optimal, given environmental externalities, a mandate is typically superior to a tax credit because the mandate results in lower levels of total fuel consumption, GHG emissions and miles traveled for the same quantity of ethanol. Because the mandates are usually implemented in tandem with tax credits (tax exemptions) and other subsidies, it is crucial to understand that subsidies in this situation reduce fuel prices, thus increasing fuel (and hence gasoline) consumption and therefore negating the stated policy goals of energy security, improved environment and farm incomes. Lapan and Moschini (2012) show that welfare, the optimal second-best mandate (expressed as a combination of a fuel tax and ethanol subsidy), dominates the optimal second-best subsidy alone.

Cui et al. (2011) show that biofuels increase welfare by increasing a country's international terms of trade in the crop exported and crude oil imported, like in the United States. Biofuels causes higher corn prices and lower crude oil prices; hence the improvements in the terms of trade are huge at about \$5.7 billion. Tax revenues decline as the ethanol replaces some gasoline, so fuel tax revenues decline and the tax credit is an added taxpayer cost. However, also important from a welfare perspective is the effect of biofuel policies on the environment. Most often the addition of biofuels increases fuel consumption, and hence increases externalities associated with vehicle miles traveled which are far greater than those due to GHG emissions. Yet these externalities are not often considered in the public debate or in economic quantitative analyses of the welfare impacts of biofuel policies. The data reported in Parry et al. (2007) imply that the share of GHG external costs per gallon of fuel represents only 2.8 percent of the total external cost of fuel consumption (associated with congestion, accidents, local pollution, and oil dependency).

Before we quantify the welfare importance of externalities related to miles traveled, let us first see how sensitive the welfare effects of biofuel policies are to the choice of a binding biofuel policy (i.e. the choice of the policy that determines the market price). In their recent welfare analysis of U.S. ethanol policies, Cui et al. (2011) assume that the U.S. corn-ethanol tax credit was a binding policy in the United States in 2009. They concluded this presumably because observed ethanol production was greater than the mandated amount, but this may occur for many reasons while still having the mandate bind for most of the year (de Gorter and Just 2010a). To make the tax credit binding, Cui et al. (2011) had to add 36 cents per gallon to the observed gasoline price in their calibration. On the other hand, de Gorter and Just (2010a) show how the mandate, coupled with the blender's tax credit, was binding. Table 6.2 presents some welfare indicators related to each method of calibration. A blend mandate determining the ethanol market price gives rise to a rectangular deadweight cost of $2.6 billion in addition to classical triangular deadweight costs. Rectangular deadweight costs refer to the quantity of ethanol production times the gap between the intercept of the ethanol supply curve and the lower free market ethanol market price.

Miles traveled externalities (–$8.2 billion), not modeled by Cui et al. (2011), are the single most important source of the overall negative net benefits of introducing corn ethanol if the mandate is binding. This points to the sensitivity of the welfare analysis of biofuel policies to the determination of the binding policy and the inclusion of important negative environmental externalities. It should be noted that the welfare loss would be even bigger if other policies present in reality, such as corn and ethanol production subsidies, were implemented along with the binding ethanol mandate (see Drabik 2011).

Table 6.2 Comparison of Welfare Effects of Ethanol Policies Relative to a No-Policy Benchmark (based on Cui et al. 2011)

	Calibrate baseline model to	
	Tax credit	*Mandate (& tax credit)*
Benefits	bil. $	bil. $
Terms of trade	5.7	5.6
Net fuel tax revenues	–2.2	–2.2
Costs		
Triangular DWCs	–1.7	–1.7
Rectangular DWC	0.0	–2.6
Costs of externalities		
GHG[a]	0.4	–0.3
Miles traveled[b]	–	–8.2
Total net benefits	6.2	–4.5

Source: calculated.

[a] Includes domestic GHGs only (second column includes byproducts and international GHGs.

[b] Congestion, accidents, pollution.

Finally, biofuel sustainability standards have become a big controversy. For example, U.S. corn-ethanol is required to reduce GHG emissions by 20 percent relative to the gasoline it is assumed to replace. These sustainability standards have a binary character, meaning that a biofuel either meets a given threshold, in which case it can be counted toward the mandate, or it does not meet it and cannot be counted. These standards are based on "life-cycle accounting" (LCA), a "well to wheel" measure of GHG emissions in the production of gasoline, and a "field to fuel tank" measure for ethanol production. The problem with this policy measure is that it does not take into account market "leakages" that may offset the implied savings measured by LCA. Two examples come to mind. First, higher corn (or oilseeds) prices means more grassland and forests will be converted into cropland, causing huge upfront GHG emissions (known as the indirect land use change effect). Second, higher biofuels production lowers crude oil prices and so other sectors and countries consume more oil. Research (e.g. Drabik and de Gorter 2013) shows that the latter leakage is very likely to prevent U.S. corn-ethanol from meeting the sustainability threshold, independent of whether the indirect land use leakage is included or not.

Whether or not consumption of biofuels results in an increase in global GHG emissions depends on two factors. First, it depends on the actual rate by which a biofuel replaces a fossil fuel. The recent literature has shown that this replacement is not one-to-one as assumed by the U.S. Environmental Protection Agency (Drabik and de Gorter 2013; Rajagopal and Plevin 2013). Second, it also depends on how much cleaner a biofuel is relative to the fossil fuel it is supposed to replace (taking into account all effects except for the less than one-to-one replacement). Interestingly, global GHG emissions could decrease due to biofuel policies even if the biofuel does not replace the fossil fuel one-to-one, provided that the biofuel has significantly lower GHG emissions than the fossil fuel. Empirically, Drabik and de Gorter (2013) find that one gasoline energy-equivalent gallon of ethanol could emit as much as 16 percent more carbon than one gallon of gasoline (under certain assumptions on the underlying parameters).

Conclusions

Because of the broad spectrum of policy goals biofuel policies are to achieve and the extensive array of biofuel policy instruments used, it is perhaps inevitable that biofuel policies have not met the policy goals set out by policymakers. We emphasize in this chapter that biofuel polices had a great impact on the world's food commodity prices, with significant social impacts on poor consumers in developing countries. We showed how the econometrics literature fails to understand the complex relationships between crop, biofuel and energy prices and all of the factors affecting these relationships, including causality and expected sign between any pairing of prices. We also showed that biofuel policies mostly fail to achieve environmental goals, especially if one includes the social costs of miles traveled due to traffic congestion and accidents. So much is yet to be done before first generation biofuels are a positive force in society.

The construction of biofuel production facilities has created short-term jobs in rural communities, but biofuel production is very capital intensive and so it requires few long-term laborers. Some studies warn that biofuels' net contribution to rural development has been very small (Swenson 2012; Low and Isserman 2009). These analyses do not take into account the fact that biofuel policy is a double-edged sword in trying to improve farm incomes and rural development. The higher crop prices due to biofuels are an implicit tax on value-added agriculture (e.g. livestock, dairy and poultry). This reduces the income of these farmers and has an impact on local economic activity. Meat and poultry production constitutes a significant amount of economic activity beyond the basic production and wholesale value stage, unlike ethanol. Elam (2012) argues that meat and poultry contribute to about 11.5 million direct-employment jobs in food processing, retailing and food service. The net impact of biofuels on rural areas is therefore unknown and may very well be negative.

Finally, a purported benefit of biofuels has been to reduce the taxpayer costs of crop subsidy programs. After October 2006 when biofuels had their first impact on grain/oilseed prices, loan deficiency and countercyclical subsidies both declined because they depend on the level of market prices. However, politicians have reacted by proposing a vastly reformed Farm Bill which is currently under consideration, whereby direct payments are eliminated and the countercyclical subsidy program is replaced by a more pernicious program that not only has higher target prices but also provides an incentive for producers to plant for government program payments instead of market signals, which will distort production decisions. Meanwhile, "revenue insurance" programs are introduced that cover "shallow losses" of crop insurance, which have become very costly to taxpayers in this high price era.

Most of the so-called reforms are largely cynical, because in this new era of high prices and record farm incomes, direct payments (where most of the "cuts" fall) were politically unviable and likely to end anyway, and countercyclical and loan deficiency payments (which are paid out when prices are low) are unlikely to be triggered. These distortions have implications for crop production, commodity prices, federal outlays, the environment, and U.S. international trade commitments.

Notes

1 For evidence, see discussion of Tables 1–3 in de Gorter et al. (2013c).

2 We define a "mandate" as being a situation where consumers have no choice to choose ethanol on the basis of miles traveled. This lack of choice and hence the ethanol price premium above that generated by the tax credit alone can be a result of several situations, including: the formal mandate itself; *de facto*

mandates in the form of environmental regulations; when consumers are unable to choose between ethanol and gasoline at the gasoline pump on a miles obtained basis because of a lack of flex cars and E-85 stations, so ethanol may be sold at a premium; or when prices are determined in world markets where biofuel policies in the rest of the world determine the ethanol price premium.

3 Discussion of Figure 6.4 gives the entire story of how corn and crude oil prices are linked, depending on which biofuel policy is binding, and contingent upon a complex web of multiple market interconnections.

4 Most economists do not analyze the price link, instead analyzing biofuels as a quantity shock. Most often, it was also no surprise:

> *"…crop diversion [for biofuel] can hardly have come as a surprise in 2006 … usage started no later than 2004 … clearly foreseeable … oilseeds for biofuel use … no surprise"*
>
> Wright (2008).

5 Using cash prices, Abbott et al. (2008, 2009) argue the multiplier is less than 2 and Mallory et al. (2012) argue it is equal to 2. The crop-biofuel price link holds regardless of what biofuel policy is binding and only diverges if there are market imperfections like ethanol blending capacity constraints.

6 The most notable, in alphabetical order, include Abbott et al. (2008, 2009), Baffes and Haniotis (2010), Carter et al. (2011), FAO (2008), Headey and Fan (2010), Hochman et al. (2011), McCalla (2009), Roberts and Schlenker (2013), Stoeckel (2008), Timmer (2010), Trostle (2008) and Wright (2011).

7 This *law* is explicitly assumed in Roberts and Schlenker (2013) where they add the calories-weighted quantity of corn, wheat rice and soybeans into one commodity. We provide empirical evidence in de Gorter et al. (2013c).

8 The free market ethanol price is determined by Equation (1) when $t_c = 0$.

9 This concept has perhaps confused readers over the years, but it is a very basic one: think of the textbook example where the demand curve for TVs on refrigerator doors does not intersect the supply curve (which, by the way, is not a reality any longer in this modern era).

10 One cannot assume that the counterfactual is the good old days when crop and biofuel prices were not linked. Although the literature does analyze only this, it is now purely academic.

11 P_{NE} adjusts for 'water' in the ethanol policy price premium.

12 The premium over the tax credit does not allow for "negative mandate premiums" (negative values reflect ethanol blending capacity constraints or ethanol production constraints or, perhaps, just measurement error) and predicted corn prices with the tax credit are not allowed to fall below the non-ethanol corn price P_{NE}.

13 Note, however, that if the shocks were such that supply increased and demand contracted, then the mandate premium could disappear, in which case the tax credit would become binding.

14 Average value in 2010/11.

15 This represents 79 percent of the total \$519 billion loss in consumer surplus due to the price change reported in Roberts and Schlenker (2013).

16 Zilberman et al. (2013) summarizes this time-series econometric literature on biofuel-crop price links by stating "… biofuel prices do not seem to affect food commodity prices…". Zhang et al. (2010, p. 445) state "Results indicate no direct long-run price relations between fuel and agricultural commodity prices, and limited if any direct short-run relationships. In terms of short-run price movements, sugar prices are influencing all the other agricultural commodity prices except rice." Sugar prices remained flat through 2007–08 whereas grain and oilseed prices went up over three times.

17 Our statistical analysis shows that ethanol prices have gone below what the minimum ethanol price should have been if the tax credit was binding—this necessarily means there were ethanol blending constraints which were expected when ethanol production was growing exponentially in the past. Note that ethanol blending constraints may also have occurred when the mandate was binding before and after the constraint, but there is no way one can identify this with statistical analysis.

18 This is not the case for sugarcane-ethanol in Brazil.

19 We use the case of corn-ethanol only as an example; the analysis is applicable for any biofuel or feedstock.

20 Of course, this assumes that *changes in* biofuel production due to a shock will have little impact on crude oil prices. Otherwise, the analysis will have to be adjusted. However, even total biofuel production has a small impact on the oil price, especially if one incorporates Hotelling's theory of exhaustible resources.

21 The ethanol price floats up and away from its lowest possible value defined in Equation (1) when linked to the gasoline (and hence oil) price through the tax credit.

22 An increase in oil prices causes gasoline prices to increase, but ethanol prices decline as do corn prices (Pathway A to B^m to C^m). Ethanol production costs go up, but have two effects in reducing corn prices: a lower ethanol price lowers corn prices (Pathway D^m), while a processing cost increase c_o causes a decrease in corn prices (Pathway C^m). Corn production costs go up, shifting the corn supply curve up and directly increasing corn prices (Pathway E^m) but this also results in a higher ethanol price, dragging corn prices further up, as per Equation (2), thereby causing a synergy effect.

23 See de Gorter et al. (2013a) for an analysis.

References

Abbott, P.C. 2012. "Biofuels, Binding Constraints and Agricultural Commodity Price Volatility". Presented at the NBER conference on Economics of Food Price Volatility, Seattle, WA, August 16. Online. Available at: http://nber.org/chapters/c12808.pdf (accessed 12 October 2013).

Abbott, P., C. Hurt, and W.E. Tyner. 2008. "What's Driving Food Prices?", Farm Foundation, Issue Report. Farm Foundation, Oak Brook, IL.

—— 2009. "What's Driving Food Prices?", March 2009 Update, Issue Report, Farm Foundation, Oak Brook, IL.

Agra Europe. 2013. "Over 60 Countries now have Biofuel Policies", August 27, 2013 No. 2580.

Baffes, J. and T. Haniotis. 2010. "Placing the 2006/08 Commodity Price Boom into Perspective", Policy Research Working Paper 5371, The World Bank, July.

Carter, C., G.C. Rausser, and A. Smith. 2011. "Commodity Booms and Busts", *Annual Review of Resource Economics* 3: 87–118.

Cui, J., H. Lapan, G. Moschini, and J. Cooper. 2011. "Welfare Impacts of Alternative Biofuel and Energy Policies", *American Journal of Agricultural Economics* 93, 5: 1235–1256.

de Gorter, H., D. Drabik, and G.R. Timilsina. 2013a. "The Effect of Biodiesel Policies on World Oilseed Markets and Developing Countries", Policy Research Working Paper #6453, The World Bank Development Research Group Environment and Energy Team, May. Available at: http://papers.ssrn.com/sol3/papers.cfm?abstract_id=2267696 (accessed 20 September 2013).

de Gorter, H., D. Drabik, E.M. Kliauga, and G.R. Timilsina. 2013b. "An Economic Model of Brazil's Ethanol-Sugar Markets and Impacts of Fuel Policies", Policy Research Working Paper #6524, The World Bank Development Research Group Environment and Energy Team, June. Available at: http://econ.worldbank.org/external/default/main?pagePK=64165259&piPK=64165421&theSitePK=469382&menuPK=64216926&entityID=000158349_20130628113935 (accessed 20 September 2013).

de Gorter, H., D. Drabik, and D.R. Just. 2013c. "Biofuel Policies and Food Grain Commodity Prices 2006–2012: All Boom and No Bust?", *AgBioForum* 16, 1: 1–13.

de Gorter, H., D. Drabik, and D.R. Just. 2013d. "The Perverse Effects of Biofuel Public-Sector Policies", *The Annual Review of Resource Economics*. 5: 463–483.

de Gorter, H. and D.R. Just. 2008. "'Water' in the U.S. Ethanol Tax Credit and Mandate: Implications for Rectangular Deadweight Costs and the Corn-Oil Price Relationship", *Applied Economic Perspectives and Policy* 30, 3: 397–410.

—— 2009a. "The Welfare Economics of a Biofuel Tax Credit and the Interaction Effects with Price Contingent Farm Subsidies", *American Journal of Agricultural Economics* 91, 2: 477–478.

—— 2009b. "The Economics of a Blend Mandate for Biofuels", *American Journal of Agricultural Economics* 91, 3: 738–750.

—— 2010a. "The Social Costs and Benefits of Biofuels: The Intersection of Environmental, Energy and Agricultural Policy", *Applied Economic Perspectives and Policy* 32, 1: 4–32.

—— 2010b. "The Social Costs and Benefits of Biofuel Policies with Pre-Existing Distortions", in G. Metcalf (ed.) *U.S. Energy Tax Policy*, Cambridge: Cambridge University Press.

Drabik, D. 2011. "The Theory of Biofuel Policy and Food Grain Prices", Charles H. Dyson School of Applied Economics and Management working paper # 2011–20, Cornell University, 12 December. Available at: http://dyson.cornell.edu/research/researchpdf/wp/2011/Cornell-Dyson-wp1120.pdf (accessed 15 October 2013).

Drabik, D. and H. de Gorter. 2013. "Emissions from Indirect Land Use Change: Do They Matter with Fuel Market Leakages?", *Review of Agricultural and Applied Economics* 16, 2: 3–15.

Du, X. and D.J. Hayes. 2009. "The Impact of Ethanol Production on US and Regional Gasoline Markets", *Energy Policy* 37, 8: 3227–3234.

Elam, T.E. 2012. "Ethanol Production: Economic Impact on Meat and Poultry Consumption, Value, and Jobs", FarmEcon LLC October 30. Available at: www.farmecon.com/Documents/RFS%20Meat%20production%20impacts%20ELAM%2010-30-12.pdf (accessed 15 October 2013).

Food and Agriculture Organization (FAO). 2008. "Soaring Food Prices: Facts, Perspectives, Impacts and Actions Required", document presented at High-Level Conference on World Food Security: The Challenges of Climate Change and Bioenergy, Rome, 3 to 5 June 2008.

—— 2013. "Biofuels and Food Security", A report by the High Level Panel of Experts on Food Security and Nutrition of the Committee on World Food Security, Rome.

Headey, D. and S. Fan. 2010. "Reflections on the Global Food Crisis: How Did It Happen? How Has it Hurt? And How Can We Prevent the Next One?", IFPRI Research Monograph 165. Washington, D.C. International Food Policy Research Institute (IFPRI).

Hochman, G., D. Rajagopal, G. Timilsina, and D. Zilberman. 2011. "The Role of Inventory Adjustments in Quantifying Factors Causing Food Price Inflation", Policy Research Working Paper 5744, The World Bank, Washington, DC, August.

Khanna, M., X. Chen, H. Huang, and H. Onal. 2011. "Land Use and Greenhouse Gas Mitigation Effects of Biofuel", *University of Illinois Law Review* 2011, 2: 549–588.

Knittel, C.R. and A. Smith. Forthcoming. "Ethanol Production and Gasoline Prices: A Spurious Correlation", *The Energy Journal.*

Kristoufek, L., K. Janda, and D. Zilberman. 2013. "Non-linear Price Transmission between Biofuels, Fuels and Food Commodities", Working Paper Series 481(ISSN 1211–3298) CERGE-EI Prague, February.

Lapan, H. and G. Moschini. 2012. "Second-best Biofuel Policies and the Welfare Effects of Quantity Mandates and Subsidies", *Journal of Environmental Economics and Management* 63, 2: 224–241.

Low, S.A. and A.M. Isserman. 2009. "Ethanol and the Local Economy: Industry Trends, Location Factors, Economic Impacts, and Risks", *Economic Development Quarterly* 23, 1: 71–88, February.

Mallory, M.L., S.H. Irwin, and D.J. Hayes. 2012. "How Market Efficiency and the Theory of Storage Link Corn and Ethanol Markets", *Energy Economics* 34, 6: 2157–2166.

McCalla, A.F. 2009. "World Food Prices: Causes and Consequences", *Canadian Journal of Agricultural Economics* 57, 1: 23–34.

Mitchell, D. 2008. "A Note on Rising Food Prices", Policy Research Working Paper Series No. 4682, Development Prospects Group, Washington D.C.: World Bank.

Parry, I.W.H., M. Walls, and W. Harrington. 2007. "Automobile Externalities and Policies", *Journal of Economic Literature* 45, 2: 373–399.

Rajagopal, D., G. Hochman, and D. Zilberman. 2011. "Indirect Fuel Use Change (IFUC) and the Life Cycle Environmental Impact of Biofuel Policies", *Energy Policy* 39, 1: 228–233.

Rajagopal, D. and R.J. Plevin. 2013. "Implications of Market-Mediated Emissions and Uncertainty for Biofuel Policies", *Energy Policy* 56, 0: 75–82.

Rajcaniova, M., P. Ciaian, and D. Drabik. Forthcoming. "International Policies on Bioenergy and Biofuels", in V.M.V. Cruz and D.A. Dierig (eds) *Industrial Crops Breeding for Bioenergy and Bioproducts*, Springer Series "Handbook of Plant Breeding".

Rausser, G.C. and H. de Gorter. 2013. "US Policy Contributions to Food Grain Commodity Prices", Paper for UNU-Wider Workshop on The Political Economy of Food Price Policy", Cornell University, Ithaca NY, 9–12 July. Available at: http://www.wider.unu.edu/publications/working-papers/2013/en_GB/wp2013–033/ (accessed 27 September 2013).

Roberts, M.J. and W. Schlenker. 2013. "Identifying Supply and Demand Elasticities of Agricultural Commodities: Implications for the US Ethanol Mandate", *American Economic Review* 103, 6: 2265–2295.

Searchinger, T., R. Heimlich, R. Houghton, F. Dong, A. Elobeid, J. Fabiosa, S. Tokgoz, D. Hayes, and T. Yu. 2008. "Use of U.S. Croplands for Biofuels Increases Greenhouse Gases through Emissions from Land use Change", *Science* 319, 5867: 1238–1240.

Serra, T. and D. Zilberman. 2013. "Biofuel-related Price Transmission Literature: A Review", *Energy Economics* 37, May: 141–151.

Stoeckel, A. 2008. "High Food Prices: Causes, Implications and Solutions", Publication No. 08/100, Rural Industries Research and Development Corporation, Australian Government, June.

Stoft, S. 2010. "Renewable Fuel and the Global Rebound Effect", Global Energy Policy Center. Research Paper No. 10–06.

Swenson, D.A. 2012. "Estimating the Importance of the Ethanol Industry to the Iowa Economy in 2011", Available at: www.econ.iastate.edu/research/other/p14755 (accessed 13 October 2013).

Timmer, P.C. 2010. "Reflections on Food Crises Past", *Food Policy* 35, 1: 1–11.

Trostle, R. 2008. "Global Agricultural Supply and Demand: Factors Contributing to the Recent Increase in Food Commodity Prices", WRS-0801, Economic Research Service, U.S. Department of Agriculture. Available at: http://www1.eere.energy.gov/bioenergy/pdfs/global_agricultural_supply_and_demand.pdf (accessed 3 September 2013).

Wright, B.D. 2008. "Speculators, Storage and the Price of Rice", ARE Update 12 no. 2, Giannini Foundation, University of California, Davis, November/December. Available at: http://giannini.ucop.edu/media/are-update/files/articles/v12n2_4.pdf (accessed 1 October 2013).

—— 2011. "The Economics of Grain Price Volatility", *Applied Economic Perspectives and Policy* 33, 1: 32–58.

Zhang Z., L. Lohr, C. Escalante, and M. Wetzstein. 2010. "Food versus Fuel: What Do Prices Tell Us?" *Energy Policy* 38, 1: 445–451.

Zilberman, D., G. Hochman, D. Rajagopal, S. Sexton, and G.R. Timilsina. 2013. "The Impact of Biofuels on Commodity Food Prices: Assessment of Findings", *American Journal of Agricultural Economics* 95, 2: 275–281.

7 Renewable Energy in the United States

Trends, Prospects, and Implications for Rural Development

Richard S. Krannich, Peter G. Robertson, and Shawn K. Olson

Introduction

Renewable energy sources, particularly wind and solar technologies, have within the past few years shifted from being novel, relatively experimental, and seemingly futuristic options for electric power production to a major force that is altering landscapes and changing the future development prospects of rural areas across much of the United States. Rapid progress in developing more practical and efficient wind and solar generating technologies for both utility-scale and local or residential-scale applications is one key factor contributing to the recent surge in renewable energy production. That technological progress has converged with growing public and political concerns about the effects of conventional electric power facilities, especially coal-fired power plants, on air quality, greenhouse gas emissions, and global climate change, creating a major push toward increased renewable energy production worldwide and in the U.S. in particular. Indeed, a recent report by the International Energy Agency suggests that by 2016 renewable energy systems will replace natural gas as the second-largest source of electric power production worldwide (IEA 2013a).

As the shift toward increased reliance on renewable energy sources continues to progress, there will be many socio-economic as well as environmental consequences for rural America. For example, rural areas with economies tied to the legacies of coal mining and past siting patterns for major electric generating stations could experience a substantial economic downturn as growing numbers of coal-fired power plants are decommissioned in response to more stringent emissions standards and the high costs of retrofitting or replacing facilities that are in many cases already approaching the end of their anticipated useful lifespan. The increasingly cost-competitive nature of some renewable energy sources—residential-scale solar power is already at retail grid parity in a few markets in the U.S. and is projected to attain full parity between 2014 and 2017 (Farrell 2013; Shahan 2013)—will undoubtedly hasten these transformations. At the same time, siting patterns for current as well as future utility-scale wind farms and solar arrays will undoubtedly lead to a concentration of such developments in rural areas—something that is all but inevitable given the large land areas required for these facilities and the greater availability and lower cost of land in rural settings. Such siting patterns have potential to produce both opportunities and liabilities for rural areas. Possible opportunities include reduced pollution, potential for employment growth, new sources of personal income, and increased revenues for local governments. Among the potential liabilities are negative effects on viewsheds, displacement of alternative land uses, possible effects on surrounding property values, damaging effects on wildlife, ecological disturbance and fragmentation effects, and public controversy and conflict.

In the remainder of this chapter we provide a brief overview of recent and anticipated trends in renewable energy development globally and in the U.S. in particular, highlighting various forces that are simultaneously fostering and constraining further expansion of renewable energy systems. We then turn to a discussion of policy issues and options pertaining to renewable energy, and provide several recommendations for new policy directions that could help to make renewable energy a more positive contributor to rural development opportunities and outcomes.

Recent and Anticipated Trends in Renewable Energy Development

Solar and wind energy technologies have seen a decade-long trajectory of very rapid growth worldwide. Electric power generation from both photovoltaic (directly converting the rays of the sun to electricity) and concentrated solar power (using the sun to heat water to power a generating turbine), wind power (using the wind to turn a turbine), and other non-hydropower renewable generation sources more than doubled between 1990 and 2012 in the U.S. alone (EIA 2013a). Although these technologies have seen impressive rates of growth, current renewable energy production from sources other than hydropower still comprises less than 6 percent of total U.S. electric power capacity. The existing organizational and physical infrastructure that has evolved over the past century around fossil fuel and hydropower electricity generation and transmission is very large and is deeply entrenched in America's economic and political systems, making a transformation to primary reliance on renewable energy sources unlikely in the short term at least. Nevertheless, rapid recent growth in installed renewable energy capacity along with rapid technology advancements, innovative financing strategies, and broad-based public and political support suggest renewable energy development will continue to accelerate and contribute an increasingly larger share of electric power production into the foreseeable future.

While there are a number of new forms of non-hydro renewable energy production, such as geothermal and biomass, the two forms of renewable energy seeing the most growth and garnering the most attention are the technologies of wind power and solar power. While both of these technologies have seen a dramatic increase in capacity, they have had different trajectories in growth and costs. In addition, these technologies are being installed at two different scales: the residential and utility scale. The growth at these different scales is driven by related forces, but because growth at one scale potentially disrupts growth at the other, the discussion of these technologies will be divided between these two scales.

Wind Power Systems

Worldwide, wind-based power generation has increased from just 17 GW (gigawatts) of capacity in 2001 to over 283 GW in 2012 (REN21 2013). Most of that growth has occurred within just the past few years, with the period between 2007 and 2012 seeing an average annual increase of 25 percent worldwide in installed wind power capacity. The U.S. has also experienced very rapid growth in wind power development, with capacity growing from 2.5 GW in 2000 to over 60 GW by 2012 (DOE 2013). Projections suggest another 20 GW will be added by 2016 if existing tax credit policies remain in place (EIA 2013b). At present the U.S. is the country with the second largest installed wind power capacity, trailing only China (GWEC 2013).

Utility-Scale Wind Power

Growth in wind power capacity has occurred almost exclusively at the utility scale, involving the development of wind farms with clusters of large turbines. These installations often involve up to two hundred or more turbines, mounted on towers that stand two to three hundred feet tall. Wind power is considered a more mature technology than solar power, and this is reflected in the share that wind plays in total U.S. electric power capacity. As of 2012 wind power contributed 28 percent of total renewable energy production (including hydropower) in the U.S., compared to just 1 percent derived from solar energy sources (EIA 2013a).

Growth in wind power capacity has occurred in part as a result of maturation of the technology, which has increased turbine capacity while also contributing to reduced turbine costs. Turbine prices did increase during the period between 2004 and 2009, due in part to increased size and performance enhancements as well as rising production costs for labor, materials, and energy (Lantz et al. 2012). More recently turbine prices have dropped—2012 prices were 25 percent lower than in 2008, and some forecasts suggest a continued trend of declining costs over the long term. Recent years have seen increased emphasis on offshore siting, along with a continuation of the trend toward larger-capacity turbines (REN21 2013). Several massive turbines standing over six hundred feet tall and with capacities of over 7 megawatts (MW) have been installed (Enercon No

Date). The trends toward larger turbines and offshore siting should contribute to further declines in price per kilowatt that accompany economies of scale and the greater consistency of offshore winds (REN21 2013).

The maturation of wind technology has allowed wind power to become less and less dependent on government subsidies and other policies designed to encourage growth in renewable energy systems. REN21 (the renewable energy policy network) reports that "onshore wind-generated power is now cost-competitive with or cheaper than conventional power in some markets on a per kilowatt-hour basis (including some locations in Australia, India, and the United States), although new shale gas in some countries is making it more difficult for wind (and other renewables) to compete with natural gas" (REN21 2013: 51). Even as surpluses of natural gas have reduced the wholesale price for electricity, wind power may still offer important advantages as a hedge against the uncertainties of future natural gas prices (Bolinger 2013).

Small-Scale Wind Power

While most installed wind power capacity involves large utility-scale facilities, power production based on small-capacity wind turbines (usually defined as less than 100 kilowatts, or kW, per turbine) is also expanding both on- and off-grid. Small-scale wind power generation, also called "distributed" wind generation, can occur at both residential and community scales. Growth at this scale has been spurred by technological advances lowering the costs of turbines and grid-connected inverters, the uncertainty of fossil fuel prices, and government incentives (REN21 2013). Growth in small-scale wind power has been rapid—as of 2012 worldwide capacity reached 576 MW, an increase of 27 percent from 2011. The bulk of small-scale wind capacity has been clustered in China (with 225 MW of installed capacity) and the U.S. (198 MW) (WWEA 2013).

Small-scale wind power is particularly well-suited for rural areas, as winds are not blocked by tall buildings and rural residents more often have access to the land area needed to install a wind turbine tower. At present investment in small-scale wind power is a longer term strategy—as of 2011 the average cost was $6,040 per kW installed (AWEA 2012), with many systems (depending on the wind, incentives, and the size of the system) offering a payback period of about ten years. Despite high initial costs, small-scale wind power can offer long-term energy cost stability by reducing or avoiding utility charges, and for some remote areas it can offer electricity at a lower cost than connecting to the grid. For U.S. farmers the USDA supports investing in small-scale wind power by offering loan guarantees for small wind turbines through the Rural Energy for America Program (REAP).

Forecasts for small-scale wind power suggest continued strong growth. WWEA (a global wind trade group) has predicted 20 percent annual increases through at least 2015 with continuation of supportive renewable energy policies along with continued technological development and increases in the generating capacity of small turbines (WWEA 2013).

Solar Power Systems

Solar power, like wind power, has been in a period of very rapid expansion due to substantial price drops, major technological advances and government support. Global installed solar capacity increased from just 1.5 GW in 2000 to 67 GW by the end of 2011, and over the past five years solar energy production has increased by an average of 50 percent per year (IEA 2013b). Strong growth in solar power is expected to continue into the foreseeable future—the International Energy Agency (IEA) predicts that by 2018 the photovoltaic (PV) solar power capacity installed worldwide will reach 308 GW, with an additional 12 GW coming from concentrated solar power (CSP) facilities (IEA 2013c).

Solar power development has expanded at an extraordinarily rapid pace in the U.S. as well. Photovoltaic solar power production grew from 138.8 MW in the year 2000 to a total of 7.2 GW (i.e. 7200 MW) by 2012 (IEA 2010; SEIA 2012). Furthermore, solar development trends appear to be very robust in 2013—during the first quarter of 2013 a record 723 MW of new solar capacity was installed (nearly as much as was installed in all of 2010). Total U.S. solar power capacity is projected to reach 9.2 GW by 2016 (SEIA 2013a).

Utility-Scale Solar Power

Unlike wind power, the development of solar power systems involves a substantial mix of both smaller-scale and utility-scale systems. Utility-scale solar power facilities contributed 2.7 GW of U.S. solar capacity in 2012, or approximately 38 percent of all solar power capacity nationwide. Utility-scale developments expected to produce an additional 3 GW of solar power were under construction as of 2012 (REN21 2013). However, this recent period of extremely rapid growth in utility-scale solar energy development appears now to be slowing as increasing numbers of utilities approach their Renewable Portfolio Standard (RPS) targets (SEIA 2013b). There is no federal RPS, and variability in state-level RPS policies has contributed to a concentration of solar and other renewable energy development in states with policies requiring that a relatively large percentage of state electricity consumption be derived from renewable-source energy supplies, those which provide strong incentives for renewable energy production, or both. For example, California has implemented an unusually aggressive RPS requiring 33 percent of the state's electric power to be derived from renewable sources by 2020 and providing considerable state-level incentives for renewable energy installations. As a consequence, California, which leads all states with 35 percent of total U.S. solar capacity, added nearly 1 GW of solar capacity in 2012 alone (REN21 2013).

Crystalline silicon photovoltaic solar has been the most common form of solar energy for both utility-scale and small-scale applications, capturing 85 to 90 percent of the solar market (Hernández-Moro and Martínez-Duart 2013). However, numerous types of PV technologies such as thin film, multi- and single-junction and concentrated photovoltaic, as well as promising technologies such as organic, quantum dot, and dye-sensitized, are emerging with the potential to increase the efficiency and applicability of solar power systems. In addition, concentrated solar power systems that use the heat of the sun to create steam and drive a turbine are now seeing strong growth. One distinct advantage of CSP technology is its ability to store energy for later use. Unlike PV systems that can be adapted to both utility-scale and smaller-scale applications, concentrated solar power is better suited for utility-scale production, and several large CSP projects totaling 1300 MW of capacity are currently under construction in desert areas of the American southwest (REN21 2013).

As is the case with wind power, PV solar technology has been in a period of strong price declines as a result of improved manufacturing processes and increased PV cell efficiencies. The average cost (both residential and utility scale) of a completed PV system reportedly declined by 24 percent over the past year, with the average price of PV panels dropping by 60 percent since 2011 (SEIA 2013c). These declines have not come without challenges to the solar power industry, as dropping prices have strained solar system manufacturers and caused some to go bankrupt.

Small-Scale Solar Power

One key difference between wind power and solar power involves the fact that PV solar is well-suited to distributed-scale installations. Distributed-scale solar is most typically installed on residential and business rooftops, and recently the development of community-shared solar arrays (commonly called "solar gardens") has gained traction, serving anywhere from ten to fifty or more households. As a result, in the U.S. and worldwide a significant amount of solar power is being generated by customer-sited solar panel installations. As of 2012, about 3.5 GW of net-metered small-scale solar generating capacity had been installed in the U.S. (Krishanmoorthy 2013).

Unlike utility-scale installations, small-scale solar power has the advantage of not having to compete directly with fossil fuel-based electric power supplies, at least not at the wholesale price level. Rather, small-scale residential and community solar electricity has to compete only with the higher retail rates customers pay for power from the grid. With small-scale solar power rapidly approaching—and in some cases attaining parity with—the retail price of electricity in many markets (Bazilian et al. 2013), electric utilities are confronted by major changes as growing numbers of traditional power consumers become electricity producers. A recent report produced for the Edison Electric Institute called "Disruptive Challenges: Financial

Implications and Strategic Responses to a Changing Retail Electric Business" (Kind 2013: 2) highlights utility industry concerns that rooftop solar technologies could threaten and fundamentally change their business model, saying:

> The combination of new technologies, increasing costs, and changing customer-usage trends allow us to consider alternative scenarios for how the future of the electric sector may develop. Without fundamental changes to regulatory rules and recovery paradigms, one can speculate as to the adverse impact of disruptive challenges on electric utilities, investors, and access to capital, as well as the resulting impact on customers from a price and service perspective. We have the benefit of lessons learned from other industries to shift the story and move the industry in a direction that will allow for customers, investors, and the U.S. economy to benefit and prosper.

These utility industry concerns result from the fact that production of grid-connected residential solar power by just a small percentage of customers has the potential to greatly reduce profits for utilities, since residential solar produces electricity during peak load periods that are also the times of highest profit for utilities. Additionally, utilities are concerned because grid-linked solar residences still use electricity during the times without sun (at night or on cloudy days), but may not consume any electricity at times when the utility can receive the greatest profit. At the same time, residential solar users are in effect receiving a subsidy because they still use the electrical infrastructure associated with the commercial power grid, even as their reduced power consumption limits their contributions to paying for the fixed capital costs of developing and maintaining the distribution system. Utilities worry that enough households switching to solar will create an amplification effect that will produce less capital to reinvest into maintenance of the grid, leading to lower system reliability and a possible need to raise rates to make up the difference in revenue lost to small-solar users—both of which may create additional incentives for residents to switch to solar.

Key Drivers of Renewable Energy Growth

The macro issues of energy security, air pollution, and more recently carbon emissions have led to supportive policies in the US encouraging renewable energy development. These policies include production targets, creation of renewable markets, and tax credits. One of the first policies designed to support renewable energy was the *Public Utility Regulatory Policies Act, 1978* (PURPA) that required utilities to buy power from small producers (under 2 kW), fostering the development of distributed renewable technologies. While access to the grid was an important key to initially opening the market for renewable energy systems, recent growth of both distributed- and utility-scale renewables has also been driven by other supportive policies, government subsidies, evolving economics, technological innovation, and state-level mandates. A brief review of some of the most important drivers of renewable energy expansion follows.

One of the most significant trends in renewable energy has been rapid technological evolution, which has led in turn to increased efficiencies and major cost reductions. As was noted above, by 2012 wind turbine prices had declined by 25 percent compared to 2008 (REN21 2013), while the costs of PV solar panels have dropped by 60 percent since 2011 (SEIA 2013c). Advances have come from technological innovations in designs, materials, and production techniques, as well as increased economies of scale, increased competition, and supply chain and installation efficiencies that come with experience. The speed at which costs have declined has meant that many policy makers likely over-estimate the current costs, and that "outdated numbers are still widely disseminated to governments, regulators and investors" (Bazilian et al. 2013: 332), obscuring the real economic viability of renewable energy.

However, despite very substantial cost declines, electricity produced via renewable energy remains more expensive than traditionally produced energy. To make up the difference, the U.S. has offered the federal renewable energy production tax credit (PTC). This subsidy has been an especially important driver influencing the growth of wind power, offering utility-scale wind turbines 2.3 cents for each kilowatt-hour (kWh) produced. Uncertainties regarding the renewal or expiration of this credit have contributed to dramatic fluctuations in wind power development, with many projects rushed to construction in advance of possible tax credit expiration dates (Barradale 2010). The PTC is only one of a myriad of incentive programs (all with different start and end dates and different technologies covered) that are available for utility-scale renewable energy. Incentives come in many forms, including corporate tax exemptions, tax credits, grants, low-interest loan programs, and favorable depreciation standards. For example, the *Modified Accelerated Cost-Recovery System* (MACRS) + Bonus Depreciation (2008–2013) modifies the depreciation schedule for certain renewable energy/energy efficiency investments, while the *Business Energy Investment Tax Credit* (ITC) offers 30 percent tax credit for solar, fuel cells, distributed-scale wind, and PTC-eligible technologies.

While the federal government has been a major subsidizer of renewable energy expansion, supportive state-level policies have also played key roles. Most states have enacted mandates or targets for renewable energy utilization, called renewable portfolio standards (RPS) or renewable electricity standards (RES), encouraging or requiring that a certain percentage of electric power must be derived from non-hydro renewable sources. These renewable portfolio standards vary widely across the twenty-nine states (along with the District of Columbia, Puerto Rico and the North Mariana Islands) that have enacted them, ranging from a mandated

33 percent renewable requirement by 2020 in California to a voluntary, unenforceable goal of 10 percent renewable power in North Dakota (DSIRE 2013). The wide-ranging nature of these standards is accompanied by considerable variation in the development of renewables—states with the strongest RPS standards have shown how quickly those mandates can spur development, while other states with high potential but without a strong RPS have seen much slower growth in renewables (EIA 2012a; Sarzynski et al. 2012). Another supportive policy often linked to RPS is the creation of renewable energy credits (REC). REC policies allow producers of renewable energy to sell "credits" linked to renewable production, in addition to the sale of electric power. One consequence of these policies has been an increase in the extent to which renewable energy generated in states with lower or less binding RPS is purchased and imported into states with a high RPS. RECs have helped to create a market for renewable energy by giving customers who do not have access to locally produced renewable electricity (or utilities that need to meet a state's standard) a means to purchase it through RECs.

Another set of policies implemented by some states (or even individual utilities) that encourage renewable growth are those involving net metering and feed-in tariffs. Net metering allows small producers to be paid or credited for the electricity they produce—if their system produces excess electricity that can be fed into the grid during the day, that unused "surplus" is credited against their use of utility-produced power at night (or over some other time frame, as some net metering policies extend over the year so, for example, customers with solar systems can receive credit for increased electrical production in the summer to offset decreased production in the winter). The number of participants in net metering has increased rapidly in recent years, and states with favorable net metering policies that allow small-scale producers to sell excess power at retail rates have experienced higher levels of adoption of small-scale renewables (EIA 2012b). Net metering offers the advantage of having been approved by the Federal Energy Regulatory Commission (FERC) and is legally tenable after having survived judicial review (Powers 2012).

Feed-in tariffs (FITs) function similarly to net metering, but go a step further by guaranteeing a price for excess energy produced that is higher than retail rates. FITs function more like the production tax credit, and are designed to guarantee a return on investment to address the high initial costs of purchasing and installing renewable systems. Although FITs have been adopted less frequently than net metering in the US, where instituted they have fostered investment in renewable systems (EIA 2013c). Their effectiveness in increasing distributed energy use in places like Germany has stimulated increased interest in this policy alternative (Couture and Cory 2009). Although the Federal Energy Regulatory Commission (FERC) has ruled that FITs are possible (FERC 2010), complex regulatory issues related to the levels at which "avoided costs" (the cost per kilowatt-hour the utility did not have to spend to produce the electricity, including the capital costs of building increased capacity and transmission) remain unresolved, a situation that at least for the present is limiting the increased use of FITs.

Federal regulations like the *Clean Air Act* have also encouraged the development of renewable energy systems by requiring reductions in emissions from conventional energy production facilities, leading to increased costs associated with the need to implement more effective pollution controls. Further pressure to reduce pollution from fossil fuel-based power production seems likely following a recent ruling by the United States Court of Appeals (District of Columbia) that the Environmental Protection Agency (EPA) has legal authority to regulate carbon emissions because they pose a danger to public health. Although the EPA has no plan currently in place to regulate carbon emissions, President Obama recently announced that the Agency would soon establish carbon pollution standards. Increasingly restrictive emissions standards and the EPA's ability to impose new mandates and penalties on carbon emissions will likely lead to increased costs for conventionally produced electric power, helping to increase the cost competitiveness of renewable energy alternatives.

Given substantial capital equipment and installation costs, financing is also key to renewable energy development. Until very recently traditional lending sources have generally viewed renewables as risky investments and been reluctant to provide favorable lending terms, leaving foreign banks to provide much of the financing for renewable projects (Mintz Levin et al. 2012). As renewable technologies have become less novel, lenders have become more willing to finance these projects, a trend that is likely to continue as renewables experience further price declines and tax credits remain stable. Large banks like Wells Fargo and JP Morgan have recently expanded their investments in distributed solar energy (Pentland 2013). In addition, some solar companies are raising money themselves by selling shares through public stock offerings (First Solar 2013), and a few traditional energy providers are also investing in renewables (Sweet 2013). New and improved financing options are likely to lower costs for renewable projects and serve as an even more important driver of renewable energy growth in the future (Mendelsohn and Feldman 2013).

While traditional financing opportunities are beginning to open up, there have also been many creative responses to the difficulty of financing the high costs of renewable systems, especially at the small scale. One example involves the formation of energy co-ops that work as a group to buy solar or wind systems, using a larger number of buyers to bargain for reduced small-scale system costs or to raise capital to purchase shares of a larger-scale renewable system, such as a solar garden. "Crowd-funding," an approach used by organizations such as the company Solar Mosaic, is another recent development. Crowd-funding involves loans provided by multiple small investors (each investing as little as $25) to provide low-rate financing of solar projects for non-profits that cannot take advantage of corporate tax incentives. Another response has involved leasing renewable systems. Residential solar customers (usually in high incentive states) enter into a leasing arrangement where solar panels are installed at very low or no cost, and then pay a locked in rate for the electricity they use. The leasing company assumes that electricity prices will continue to increase, and expects to profit by selling excess power produced by the system at increasing rates while offering stable prices for the homeowner.

Another recent financing strategy has been the creation of Property Assessed Clean Energy (PACE). Designed to encourage residential and other distributed-scale renewable energy systems or energy efficiency upgrades, PACE financing is like a loan that spreads out the high initial cost of these systems over a longer time period. PACE allows governments to create financing districts that offer loans for the cost of renewable systems. The loan value is added as an increased assessment on the property of owners who install renewable energy, and the debt is repaid through increased property taxes. This increases yearly property tax but gives property owners the assurance they can recoup the investment even if they sell the property, because the value of the upgrade will be included in the property taxes the next owner will pay.

In addition to technological changes, supportive public policies, and increased financing availability, broad-based public support has also been an important driver of renewable energy expansion. For example, a March 2013 Gallup poll asked "Do you think that as a country, the United States should put more emphasis, less emphasis, or about the same emphasis as it does now on producing domestic energy from each of the following sources?" Three out of four (76 percent) of the respondents indicated they wanted more solar and 71 percent wanted more wind power, compared to just 31 percent wanting more use of coal, 37 percent wanting more nuclear power, and 46 percent wanting more use of oil (Gallup Politics 2013). Such findings are consistent with those derived from many other national and regional opinion polls and surveys that have found high levels of public support for renewable technology since the technology was first commercialized in the late 1970s (Brunner and Vivian 1980). Even in states like Montana, Wyoming, and Utah that have long traditions of reliance on extractive economies, public support for renewable energy is high (Colorado College State of the Rockies Report 2013). Further, public recognition of the adverse environmental consequences of existing technology (but not concern about carbon emissions or relationships to global warming) has been found to be more instrumental in shaping attitudes about renewable energy than the costs of shifting to these new technologies (Ansolabehere and Konisky 2012). Furthermore, even as the costs of renewable energy move closer to those associated with conventional sources, most consumers indicate they are willing to pay more for renewable energy; a 2011 study found the average U.S. citizen is willing to pay an additional $162 annually for power from renewable systems (Aldy, Kotchen and Leiserowitz 2012).

Finally, it is important to note that public support for utility-scale renewable energy development often forms locally in response to expectations about the economic benefits projects may bring to an area in the form of new tax revenues, new employment opportunities, and new sources of income or reduced electricity costs (Bidwell 2013; Slattery et al. 2012). Most revenue derived from utility-scale projects comes from taxes assessed on the value of installed facilities and equipment. Given the high cost of turbines and solar panels this can be a substantial source of revenue for a rural county, especially when large projects are sited in sparsely populated areas. Also, unlike many other types of economic development in rural areas that tend to attract substantial workforce in-migration, these projects typically do not require significant public expenditures on services such as additional roads, sewers, police, or schools. However, because of a rapid depreciation rate applied to taxable capital equipment at renewable energy facilities, what may be a large initial revenue increase for local governments is often relatively short-lived. Some states have implemented additional revenue-generating policies, which provide longer-lived economic benefits. For example, in 2010 the state of Wyoming implemented a Wind Energy Excise Tax, which in 2012 began collecting $1/megawatt-hour annually from developers for all wind electricity generated in the state; 60 percent of this revenue goes directly to the county hosting the facility.

Utility-scale wind farms and solar arrays can create much-needed employment and income opportunities in areas where such facilities tend to be sited. In a study examining the economic effects of wind power development from 2000 to 2008, Browne et al. (2012) reported an aggregate increase in county-level personal income of approximately $11,000 and the addition of 0.5 new jobs for each megawatt of installed wind power capacity. Most project-induced jobs involve short-term construction-phase employment, as renewable projects typically require only a small number of employees during the operation phase. Also, there is a tendency for relatively few of these newly-created jobs to be filled by established local-area residents. Nevertheless, even a handful of new jobs can represent a substantial economic boost in many rural areas, as can short-term construction-related increases in sales for local-area materials providers and service-sector businesses.

In areas where development occurs exclusively or extensively on privately owned land, utility-scale renewable energy facilities can also produce an economic windfall for landowners who lease their property for the siting of wind turbines or solar arrays. Lease payments can be very helpful for rural landowners as they provide a steady and predictable income, while often not displacing revenue generated through farming or other existing land uses. Additionally, some wind developers have begun offering "good neighbor payments" or "community benefit provisions" to households living adjacent to turbine-leased land, since those neighbors often feel impacted by the installation of turbines. In some counties, economic benefits from wind farms are distributed even more widely. For example, in Sherman County, Oregon a compensation program provides an annual royalty payment check to every household in the county, enabling all residents to share in the revenue stream from the wind farm (Druckenmiller 2012). Even in areas without utility-scale growth, distributed or small-scale projects can provide economic stability for businesses and residents that install systems as a buffer against rising prices and as a long-term investment. For some rural residents in very remote areas, renewable systems can provide electricity much cheaper than connecting to the larger grid.

Barriers and Liabilities

Although a variety of forces are driving the growth of renewable energy capacity in the U.S., there are also multiple factors acting to constrain that expansion. Among the more important of these are pricing and cost issues. At present conventional electric power sources maintain a considerable price advantage over renewable energy in most markets, due in part to the fact that many of the negative externalities associated with conventional power systems are not reflected in the price of electricity. Rather, conventional energy production is partially subsidized because the human health and environmental quality effects of power plant emissions and fossil fuel extraction are not fully reflected in the price of electricity. Also, while the costs of both utility-scale and distributed-scale renewable energy options are declining, most renewable power options remain more expensive than grid-supplied electricity, due in part to the still-emergent nature of these technologies and the limited availability of financing.

The complexities of the regulatory environment involving electric power production and distribution further exacerbate the cost disadvantages that constrain renewable energy expansion. The U.S. electric utility industry is highly regulated, with multiple layers of regulation at federal, state, municipal, and utility-sector levels. The resulting complexity and variability of regulations across multiple layers of authority and from one geographic location to another can add considerably to the cost of renewable energy installations. Indeed, one analyst has asserted that "up to 40 percent of the cost of installing solar panels onto your home or business isn't related to hardware at all, but rather due to complications of 'soft costs,' like permitting, zoning, and hooking your system up to the power grid" (Simmons 2011). Furthermore, "inconsistencies in soft cost requirements—from town to town and utility to utility—make it difficult for solar installers to enter new markets" (Le 2013).

There are also important barriers associated with the difficulties involved in integrating renewable energy production into the existing electric power grid and transmission systems. Provision of a sufficient and reliable supply of electricity requires a perfect balancing of supply with demand. Renewable energy sources present a challenge to attaining that balance, since the amount of power produced can vary from one minute to the next as wind speed variations and changes in cloud cover alter output from wind turbines and solar panels. Substantial variations in output from renewable energy sources also occur from day to night and from season to season. This variability requires that utilities have back-up capacity available and ready at a moment's notice to fill in the troughs in power supply caused by shifts in renewable output or in power demand. Some have argued that as renewable energy production becomes more widespread these fluctuations will level out as a result of anticipated advances in power storage technologies and a greater geographic dispersion of renewable systems, allowing dips in output from one area to be counterbalanced by sustained or increased output from others (Grossmann et al. 2012). However, such "balancing" outcomes will require substantial changes to the national power grid, a highly-complex system that at present provides for only limited inter-regional interconnection and coordination (Kassakian et al. 2011).

Further complicating the situation are difficulties involving the ability to transmit power from remotely sited utility-scale renewable energy developments to distant power consumption markets. Siting requirements and the renewable resource conditions needed for utility-scale wind farms and solar arrays tend to concentrate such developments in remote rural areas, often at considerable distance from population centers that are the biggest users of electric power. The absence or inadequacy of existing power transmission lines capable of moving this newly-generated electricity across long distances means that additional high-capacity transmission lines are needed. Adding the needed transmission capacity can be difficult to achieve given high construction costs and the challenges associated with the permitting of transmission corridors, which are often highly controversial due to public concerns about environmental and land use consequences (Brown and Rossi 2010). The need to address the requirements of multiple levels of regulatory authority makes transmission lines difficult to site, and the barriers associated with public opposition and reluctance to cooperate by what may be hundreds or thousands of individual landowners are daunting (Furby et al. 1988). Policy responses to these difficulties, such as the creation of National Interest Corridors that assign the Federal Energy Regulatory Commission final authority over interstate transmission decisions, could perhaps facilitate future siting efforts (McLaughlin 2008), but the imposition of federal authority in locating and permitting transmission corridors is also likely to spawn considerable controversy. This highlights another advantage of distributed energy options. With power from such systems used primarily where it is produced, the need for new high-voltage transmission lines to deliver power to distant locations is reduced.

Policy inconsistencies also act to constrain the growth of renewable energy. In particular, the availability of federal funding programs in support of renewable research and development activities has fluctuated wildly as some various renewable technologies gain or lose favor and as the politics of energy policy swing in various directions. For example, recurring Congressional threats to discontinue the production tax credit every few years have undoubtedly had a dampening effect on private-sector decisions about investment in new renewable energy projects. As Liang and Fiorino (2013: 111) has observed, "incremental, predictable, and credible expenditures appear to be more conducive to renewable energy development" but "a boom–bust cycle of resource support hardly translates policy goals into intended results."

The interests and concerns of the electric utilities industry may also serve to limit growth in renewable energy production. The prospect of increased utilization of residential-scale distributed renewable energy has spawned considerable concern within the utility industry due to the potential for declining revenues, increased costs, and lower profitability. The practice of net metering, which requires that utilities purchase excess power produced by individual small-scale and residential renewable installations, has increasingly been a focus of industry efforts to secure regulatory changes that would reduce compensation to or increase costs for the owners of those small-scale systems. Similarly, a number of utilities have pursued policy shifts that would allow them to restructure or limit their contractual obligations to purchase renewably-produced electricity from larger-scale commercial facilities. Utilities have substantial lobbying influence and legal resources, and their ongoing efforts to protect the industry from these "disruptive challenges" to the centralized utility services business model (see Kind 2013) may create new barriers to more widespread adoption of renewable energy options.

Finally, it is important to acknowledge that local-level public concerns about and opposition to the siting of large-scale renewable installations can be an important barrier to the expanded production of renewably produced energy. Despite broad-based public support for renewable energy development generally, local-area residents, landowners and land users are often strongly opposed to the development of these facilities. Indeed, a recent study indicates that many proposed wind projects in the U.S. have been blocked as a result of opposition and controversy at the local level (Pociask and Fuhr 2011). Public concerns about renewable projects vary widely, but most commonly include fears that they will spoil views, create unwanted noise, cause harm to wildlife and habitats, change and restrict land use patterns and options, or lower property values. In addition, concerns sometimes arise regarding the fact that local communities and rural areas will bear the brunt of potentially adverse effects associated with these facilities, while most of the economic benefits as well as the newly generated electricity will be sent to other, often distant urban areas.

Policy Options and Recommendations

Policies that promote the development of renewable energy at both the utility scale and smaller scales seem necessary for the foreseeable future. At the same time, policies designed to promote distributed generation may offer longer-term opportunities and benefits to rural areas, particularly since forecasts point strongly toward a diminishing importance of utility-produced electricity. To best promote renewable energy in ways that will reinforce and enhance positive rural development outcomes, we offer the following policy recommendations:

- **Establish and maintain consistent and predictable policies**
  Even as renewable technologies continue to mature, for the near term at least there is a need to continue supportive policies such as the production tax credit, the residential renewable energy tax credit, or feed-in tariffs. Such policies also need to be characterized by a higher degree of predictability, stability, and consistency than has occurred over the past several years. For example, recurring Congressional threats to discontinue the production tax credit every few years have undoubtedly had a dampening effect on private-sector decisions about investment in new renewable energy projects. Long-term policy commitments will help signal to investors and inventors that these technologies are worth pursuing. At the same time, to provide the correct incentives such policies should include a defined time horizon and a reduction schedule that will lower subsidy levels as renewable energy technologies mature and as new innovations and increased efficiencies lead to a declining need for subsidization and incentive programs.

- **Coordinated planning at the federal, regional, state, utility, and local levels**
  The electric power grid is vast and complex, and the intermittent nature of power production from renewable sources poses serious challenges to the existing grid system. To be stable and reliable the grid will require the interconnection of backup capacity and increased integration across multiple scales of generating systems. Further deployment of smart-grids and deeper interconnection of the national grid will help smooth the intermittencies of renewable energy supplies and increase the total amount of renewable capacity the grid can use. However, the successful deployment of these systems needs careful planning, with emphasis on more effectively connecting the many levels of bureaucracy involved in regulating the electrical system.

- **Low-interest loan programs for small-scale renewable systems**
  The high initial cost of a residential-scale solar or wind system prevents many would-be buyers from installing renewable energy. Although

banking practices are changing, traditionally banks have not offered favorable loans for renewable energy projects even if they can be shown to offer consistent returns of reduced energy costs. Because of this we recommend the continuation of programs like the Rural Energy for America Program (REAP) that offer low-interest loans for renewable and energy efficiency projects. Given the potential for substantial energy cost savings, and because residential-scale energy systems are usually covered by homeowners' insurance, such investments tend to be among the safest available to individual property owners and to lending institutions.

- **Local-level revenue generation standards**
Most of the public revenue derived from utility-scale wind and solar energy developments comes in the form of taxes on capital equipment installed at these facilities. Because a rapid depreciation schedule is typically applied to such equipment, it is important that counties and municipalities hosting developments plan carefully around the timing and uses of these funds. Renewable energy projects have the potential to provide much-needed tax revenues for rural areas, but the relatively short-term and declining nature of revenue flows can create challenges in anticipating the extent and duration of revenue increases and in determining how funds should be used. Given the relatively short-term and declining nature of these revenue flows, the benefits derived from such funds might be enhanced by creating investment accounts or endowments that can help to address public expenditure needs over a much longer term. State- and county-level governments could also consider implementing policies that produce a steadier income stream, such as Wyoming's Wind Excise Tax, as well as compensation programs that distribute economic benefits to a wider number of recipients, as has occurred in Sherman County, Oregon. However, it is important to weigh such options carefully since increased taxes and economic disbursements required of renewable energy developers could constrain an industry that already faces a number of vulnerabilities.

Local governments would also benefit from the creation and use of standardized agreements that clearly establish the amount of tax revenue they will receive, and how revenue levels will change over time. Standardized agreements could help local governments avoid being "bargained out of" revenues by energy companies that may negotiate for reduced taxes on the basis of job creation promises that may not materialize. In addition, a modest energy production tax allocated to local governments in areas affected by the siting of utility-scale renewable energy facilities would help to insure that some revenues remain available over the lifespan of these projects to offset longer-term externalities imposed on the locality.

- **Local job creation and hiring policies**
  Although large-scale renewable energy projects are often promoted as providing significant new job opportunities for host areas, those expectations and promises frequently give way to disappointment in the face of limited job creation and a tendency for many positions to be filled by non-local and in-migrating workers. Policies and programs that help to increase employment opportunities for existing local-area residents are needed if renewable energy projects are to contribute more effectively to rural economic development outcomes. Such policies might include minimum requirements for a percentage of employees to be hired from the surrounding area during project construction and operation periods, as well as requirements for local purchase of some specified portion of construction materials and services. In addition, establishment of training programs to enhance preparedness of local-area residents for employment at renewable facilities should be considered.

- **"Best practice" siting and interconnection standards**
  We also suggest the creation of new siting standards focused on reducing the negative externalities that are likely to accompany development of utility-scale renewable energy facilities. Besides the requisite environmental impact assessment and mitigation considerations, we suggest that standards include recognition of the visual impacts these developments can create. While the visual impact of wind turbines or other large renewable energy structures may appear unimportant compared to the environmental damage associated with some extractive industries, the visual and environmental effects of renewable technologies are not benign. Utility-scale projects require careful planning, with siting standards that include well-defined setback and distance requirements that protect cultural and natural landscapes. A careful planning process that considers and includes public views and concerns as part of the site evaluation and decision process will help to avoid or reduce the controversy and opposition that often arises in response to large-scale project proposals.

  For residential and small-scale systems, we recommend the adoption of interconnection and permit standards that facilitate and encourage the connection of these systems to the grid. Policies are needed that prevent the imposition of unnecessary permitting and connection requirements or expenses. These standards should also fully consider how residential-scale renewable energy systems offer benefits grid-wide, such as reduced utility power production and transmission costs and environmental benefits, and should ensure that owners will receive a fair price for the excess electric power they produce.

Conclusions

Although the future trajectory for renewable energy remains uncertain, both the need for an energy system with a smaller environmental footprint and current trends in the creation and use of renewable technologies strongly suggest that our future world will use more and more renewably produced electricity. As that future unfolds, rural areas will experience potentially profound changes linked to increased reliance on both utility- and residential-scale renewable energy systems. Through careful planning and policy development those changes can contribute positively to economic, environmental, and social well-being, and help to foster positive rural development outcomes.

References

Aldy, J.E., M.J. Kotchen, and A. Leiserowitz. 2012. 'Willingness to pay and political support for a US national clean energy standard', *Nature Climate Change* 2: 596–599.

Ansolabehere, S. and D.M. Konisky. 2012. 'The American public's energy choice', *Dædalus* 141, 2: 61–71.

AWEA (American Wind Energy Association). 2012. *2011 U.S. Small Wind Turbine Market Report*. Online. Available at: http://awea.files.cms-plus.com/2011_AWEA_Small_Wind_Turbine_Market_Report.pdf (accessed 29 July 2013).

Barradale, M. 2010. 'Impact of public policy uncertainty on renewable energy investment: Wind power and the production tax credit', *Energy Policy*, 38, 12: 7698–7709.

Bazilian, M., I. Onyeji, M. Liebreich, I. MacGill, J. Chase, J. Shah, D. Gielen, D. Arent, D. Landfear, and S. Zhengrong. 2013. 'Re-considering the economics of photovoltaic power', *Renewable Energy* 53: 329–338.

Bolinger, M. 2013. *Revisiting the Long-Term Hedge Value of Wind Power in an Era of Low Natural Gas Prices*. Lawrence Berkeley National Laboratory Report. Online. Available at: http://emp.lbl.gov/sites/all/files/lbnl-6103e.pdf (accessed 29 July 3013).

Bidwell, D. 2013. 'The role of values in public beliefs and attitudes towards commercial wind energy', *Energy Policy* 58: 189–199.

Brown, A.C. and J. Rossi. 2010. 'Siting transmission lines in a changed milieu: Evolving notions of the public interest in balancing state and regional considerations', *U. Colo. L. Rev.* 81: 705–770.

Brown, J.P., J. Pender, R. Wiser, E. Lantz, and B. Hoen. 2012. 'Ex post analysis of economic impacts from wind power development in US counties', *Energy Economics* 34: 1743–1754.

Brunner, R.D. and W.E. Vivian. 1980. 'Citizen viewpoints on energy policy', *Policy Sciences* 12, 2: 147–174.

Colorado College State of the Rockies. 2013. *Conservation in the West Poll 2013: Survey of the Attitudes of Voters in Six Western States*. Online. Available at: www.stateoftherockies.com (accessed 29 July 2013).

Couture, T. and K.S. Cory. 2009. *State Clean Energy Policies Analysis (SCEPA) Project: an Analysis of Renewable Energy Feed-In Tariffs in the United States*.

National Renewable Energy Laboratory. Online. Available at: www.nrel.gov/docs/fy09osti/45551.pdf (accessed 29 July 2013).

DOE (Department of Energy). 2013. Installed Wind Capacity. Online. Available at: http://www.windpoweringamerica.gov/wind_installed_capacity.asp (accessed 10 June 2013).

Druckenmiller, H. 2012. 'At wind speed: How the U.S. wind industry is rapidly growing our local economies', National Resources Defense Council, Issue Paper. Online. Available at: www.nrdc.org/energy/wind-powered-communities/files/wind-communities-ib.pdf (accessed 29 July 2013).

DSIRE (Database of State Incentives for Renewable and Efficiency). 2013. Renewable Portfolio Standard Policies. Online. Available at: www.dsireusa.org/documents/summarymaps/RPS_map.pdf (accessed 24 July 2013).

EIA (Energy Information Agency). 2012a. 'Electricity generation from non-hydro renewable sources varies by state'. Online. Available at: www.eia.gov/todayinenergy/detail.cfm?id=6090 (accessed 24 July 2013).

—— 2012b. 'Participation in electric net-metering programs increased sharply in recent years'. Online. Available at: www.eia.gov/todayinenergy/detail.cfm?id=6270 (accessed 24 July 2013).

—— 2013a. 'Energy in brief: How much of our electricity is generated from renewable energy?' Online. Available at: www.eia.gov/energy_in_brief/article/renewable_electricity.cfm (accessed 13 July 2013).

—— 2013b. 'Extended tax credits could raise projected renewable electricity capacity and generation'. Online. Available at: http://www.eia.gov/todayinenergy/detail.cfm?id=11371 (accessed 13 June 2013).

—— 2013c. 'Feed-in tariff: A policy tool encouraging deployment of renewable electricity technologies'. Online. Available at: www.eia.gov/todayinenergy/detail.cfm?id=11471 (accessed 24 July 2013).

Enercon. No date, 'Technical data'. Online. Available at: www.enercon.de/en-en/66.htm (accessed 29 July 2013).

Farrell, J. 2013. 'Is solar cheaper than grid electricity? Yes and no.' Online. Available at: www.ilsr.org/solar-cheaper-grid-electricity-yes-no/ (accessed 29 July 2013).

FERC (Federal Energy Regulatory Commission). 2010. 'FERC clarifies California feed-in tariff procedures'. Online. Available at: www.ferc.gov/media/news-releases/2010/2010–4/10–21–10–E-2.asp (accessed 29 July 2013).

First Solar. 2013. 'First Solar announces public offering of common stock'. Online. Available at: http://investor.firstsolar.com/releasedetail.cfm?releaseid=770731 (accessed 25 July 2013).

Furby, L., P. Slovic, B. Fischhoff, and R. Gregory. 1988. 'Public perceptions of electric power transmission lines', *Journal of Environmental Psychology* 8, 1: 19–43.

Gallup Politics. 2013. 'Americans want more emphasis on solar, wind, natural gas'. Online. Available at: www.gallup.com/poll/161519/americans-emphasis-solar-wind-natural-gas.aspx (accessed 25 July 2013).

Grossmann, W., I. Grossmann, and K. Steininger. 2012. 'Distributed solar electricity generation across large geographic areas, Part I: A method to optimize site selection, generation and storage', *Renewable and Sustainable Energy Reviews*.

GWEC (Global Wind Energy Council). 2013. 'Global wind statistics 2012'. Online. Available at: www.gwec.net/wp-content/uploads/2013/02/GWEC-PRstats-2012_english.pdf (accessed 24 July 2013).

Hernández-Moro, J. and J.M. Martínez-Duart. 2013. 'Analytical model for solar PV and CSP electricity costs: Present LCOE values and their future evolution', *Renewable and Sustainable Energy Reviews* 20: 119–132.

IEA (International Energy Agency). 2010. 'Trends in photovoltaic applications; Survey report of selected IEA countries between 1992 and 2009', Report IEA-PVPS T1–19.

—— 2013a. 'Renewables to surpass gas by 2016 in the global power mix'. Online. Available at: www.iea.org/newsroomandevents/pressreleases/2013/june/name,39156,en.html (accessed 29 July 2013).

—— 2013b. 'FAQs renewable energy'. Online. Available at: www.iea.org/aboutus/faqs/renewableenergy/ (accessed 29 July 2013).

—— 2013c. 'Medium-term renewable energy market report 2013: Market projections to 2018'.

Kassakian, J.G., R. Schmalensee, et al. 2011. *The Future of the Electric Grid: An Interdisciplinary MIT Study*. Technical Report, Massachusetts Institute of Technology.

Kind, P. 2013. *Disruptive Challenges: Financial Implications and Strategic Responses to a Changing Retail Electric Business*. Report for Edison Institute. Washington, DC.

Krishanmoorthy, B., M. Taylor, and B. Campbell. 2013. *2012 SEPA Utility Solar Rankings*. Report for Solar Electric Power Association.

Lantz, E., R. Wiser, and M. Hand. 2012. *The Past and Future Cost of Wind Energy*. IEA Wind Task 26. Technical Report for the National Renewable Energy Laboratory. NREL/TP-6A20–53510.

Le, M. 2013. 'Finding solutions to solar's soft cost dilemma'. Online. Available at: http://energy.gov/articles/finding-solutions-solars-soft-cost-dilemma (accessed 24 July 2013).

Liang, J. and D.J. Fiorino. 2013. 'The implications of policy stability for renewable energy innovation in the United States, 1974–2009', *Policy Studies Journal* 41, 1: 97–118.

McLaughlin, N.A. 2008. 'Condemning open space: Making way for National Interest Electric Transmission Corridors (or Not)', *Virginia Environmental Law Journal* 26: 399.

Mendelsohn, M. and D. Feldman. 2013. *Financing US Renewable Energy Projects Through Public Capital Vehicles: Qualitative and Quantitative Benefits*. National Renewable Energy Laboratory. Technical Report NREL/TP-6A20–58315.

Mintz Levin, Cohn, Ferris, Glovsky and PC, Popeo. 2012. 'Renewable energy project finance in the U.S.: 2010–2013 overview and future outlook'. Online. Available at: www.mintz.com/DesktopModules/Bring2mind/DMX/Download.aspx?EntryId=231&PortalId=0&DownloadMethod=attachment (accessed 12 July 2013).

Pentland, W. 2013. 'Another day, another $1 billion for distributed solar'. Online. Available at: www.forbes.com/sites/williampentland/2013/06/26/another-day-another-1–billion-for-distributed-solar/?ss=energy-management (accessed 19 July 2013).

Pociask, S. and J.P. Fuhr. 2011. *Progress Denied: A Study on the Potential Economic Impact of Permitting Challenges Facing Proposed Energy Projects*. Report for US Chamber of Commerce. TeleNomic Research, LLC.

Powers, M. 2012. 'Small is (still) beautiful: Designing U.S. energy policies to increase localized renewable energy generation', *Wisconsin International Law Journal* 30, 3: 595–667.

REN21. 2013. 'Renewables 2013 Global Status Report'. Online. Available at: http://www.ren21.net/Portals/0/documents/Resources/GSR/2013/GSR2013_lowres.pdf (accessed 29 July 2013).

Sarzynski, A., J. Larrieu, and G. Shrimali. 2012. 'The impact of state financial incentives on market deployment of solar technology', *Energy Policy* 46: 550–557.

SEIA (Solar Energy Industries Association). 2012. 'U.S. Solar Market Insight 2012 Year in Review'. Online. Available at: www.seia.org/research-resources/us-solar-market-insight-2012–year-review (accessed 13 June 2013).

—— 2013a. 'U.S. Solar Market Insight Q1 2013'. Online. Available at: www.seia.org/research-resources/us-solar-market-insight-q1–2013 (accessed 13 June 2013).

—— 2013b. 'U.S. Solar Market Insight Q2 2013'. Online. Available at: www.seia.org/research-resources/us-solar-market-insight-report-2013–q2 (accessed 22 October 2013).

—— 2013c. 'Solar Industry Data'. Online. Available at: www.seia.org/research-resources/solar-industry-data (accessed 13 June 2013).

Shahan, Z. 2013. 'US solar to hit grid parity 2014–2017'. Online. Available at: http://cleantechnica.com/2013/06/18/us-solar-to-hit-grid-parity-2014–2017–more-solar-pv-charts-sort-of (accessed 29 July 2013).

Simmons, G. 2011. 'Rooftop solar challenge to cut solar's red tape'. Online. Available at: energy.gov/articles/rooftop-soar-challenge-cut-solars-red-tape (accessed 29 July 2013).

Slattery, M.C., B.L. Johnson, J.A. Swofford, and M.J. Pasqualetti. 2012. 'The predominance of economic development in the support for large-scale wind farms in the U.S. Great Plains', *Renewable and Sustainable Energy Reviews* 16: 3690–3701.

Sweet, C. 2013. 'Duke Energy invests in solar firm Clean Power Finance'. Online. Available at: http://online.wsj.com/article/BT-CO-20130627–704626.html?mod (accessed 25 July 2013).

WWEA (World Wind Energy Association). 2013. '2013 Small Wind World Report Update'. Online. Available at: www.wwindea.org/webimages/SWWR_summary.pdf (accessed 29 July 2013).

8 Rural Energy Use and the Challenges for Energy Conservation and Efficiency

Matteo Muratori

Introduction

The current concerns about dependence on fossil fuels and the corresponding greenhouse gas emissions are spurring interest in the use of policy and technology solutions to curb and rationalize energy consumption and reduce U.S. dependency on fossil fuels.

Two solutions are proposed to mitigate the problems associated with a fossil fuel economy: large-scale usage of alternative and renewable energy sources, and considerable improvements in energy conservation and efficiency. Improvements in energy use (including reduced environmental impact) and conservation are an absolute necessity to assure a sustainable energy future.

Residential and transportation energy consumption results from a combination of infrastructure and behavior factors, which make rural and urban areas quite different. Residential energy consumption in rural areas is almost 10 percent higher compared to urban areas, while electricity consumption is about 50 percent higher. This indicates a difference in the way that energy is used in rural areas, especially for space heating. This different choice of energy source is partially justified by infrastructure differences, especially regarding residential use of natural gas.

The higher energy consumption in rural areas is caused by a diversity of factors, ranging from household characteristics, socio-economic dynamics, and environmental conditions. The typical image of a rural house is of an old, large farmhouse. Rural households are indeed larger than urban residences. On average, rural homes are 30 percent larger than urban homes, and they are typically detached houses, which means that they are more exposed to weather conditions and do not benefit from radiant heat from adjacent buildings. On the other hand, unlike the stereotype, rural houses are typically newer compared to their urban counterparts. Since homes built since 1990 are on average 27 percent larger than homes built in earlier decades, the differences in rural and urban energy use may be expected, since much of the residential energy use is directly correlated with the size of the home. On the other hand, newer households typically manage energy better, with greater use of insulation and better energy efficiency per square foot. Thus, the energy disparity between rural and urban households is not merely because of the home size.

Moreover, families living in rural areas drive about seven thousand more miles annually than their urban counterparts, which results in an additional 330 gallons of gasoline consumed every year (assuming an average efficiency of about 21 miles per gallon). Thus, increasing energy efficiency and conservation in both rural and urban settings requires policy incentives that respond to the different conditions and realities in these areas.

This chapter deals with energy conservation and efficiency, with particular focus on personal transportation and residential energy consumption in rural areas. The most prominent implications found are:

1. Rural households are about 30 percent bigger, but use only 10 percent more energy. Therefore, rural households are generally more energy efficient per square foot than urban households. This means that improved energy efficiencies may be more difficult to achieve than in urban households.
2. Rural households have fewer energy options. Natural gas is less common for space heating, likely because the natural gas infrastructure is not as well developed in rural areas.
3. Rural energy use is related to the rural lifestyle. Most often and in many ways, this lifestyle is a choice.
4. There are different vehicles used in rural settings than in urban settings and they are used in different ways. In rural communities, there is a greater use of larger vehicles, such as pick-up trucks. While there are many advances being made by car manufacturers in alternative transportation, these are largely in smaller passenger vehicles. Where options exist for larger vehicles, such as pick-up trucks and medium duty trucks, the lack of a rural infrastructure may make adoption more difficult (i.e. the adoption of natural gas vehicles, which are currently being proposed as a cheaper alternative to gasoline, might be more difficult in rural areas).
5. Nine out of ten households in the United States are in urban settings, making it harder to justify federal and national investments and policies directly oriented to reduce rural energy consumption.

Current Energy Situation, Residential and Personal Transportation Consumption

According to the U.S. Energy Information Administration, the total U.S. primary energy consumption totaled 78096 quadrillion BTU in 2011, an increase of approximately 4 percent from 2010. Of this primary energy, 36 percent came from petroleum, 26 percent from natural gas, 20 percent from coal, 9 percent from renewables (34 percent hydroelectric, 2.5 percent geothermal, 1.7 percent solar, 12.6 percent wind, and 48.8 percent biomass), and 8 percent from nuclear power (please see Figure 8.1 for the breakdown in energy use by source and sector).

A few important trends regarding energy in the United States are notable:

- Petroleum is the single most important source of primary energy, accounting for 36 percent of the total energy consumption.
- Natural gas is used equally to produce electricity in the industrial, residential and commercial sectors, but it is essentially unused for transportation, which is very different from other countries around the world which make far greater use of natural gas. Given the recent developments in horizontal drilling and hydraulic fracturing, natural gas is expected to play a much greater role in the U.S. energy future in the near–medium term.
- Coal is mainly used (92 percent) for electricity production.
- Nuclear power is entirely used for electricity production.

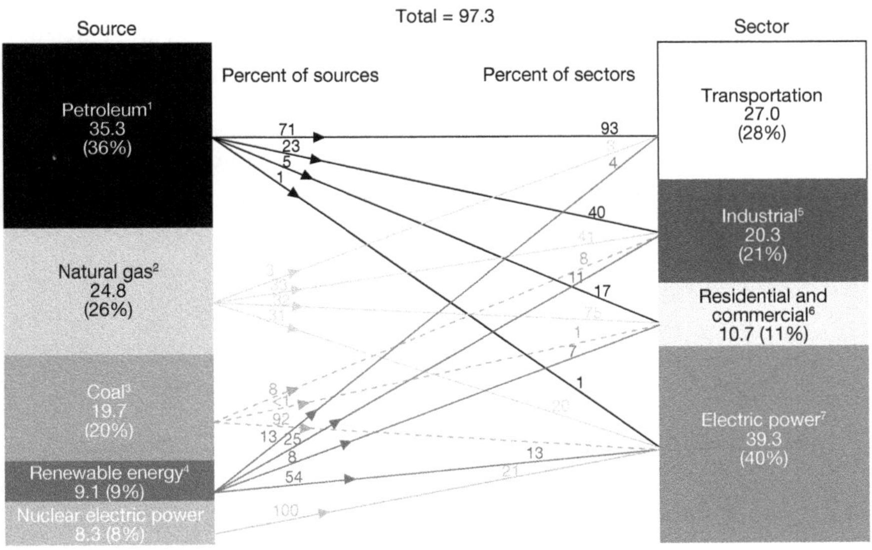

Figure 8.1 Primary Energy Consumption by Source and Sector, 2011 [Quadrillion BTU].

Source: U.S. EIA Annual Energy Review 2012.

1 Does not include biofuels that have been blended with petroleum—biofuels are included in "Renewable Energy."
2 Excludes supplemental gaseous fuels.
3 Includes less than 0.1 quadrillion Btu of coal coke net exports.
4 Conventional hydroelectric power, geothermal, solar/PV, wind, and biomass.
5 Includes industrial combined-heat-and-power (CHP) and industrial electricity-only plants.
6 Includes commercial combined-heat-and-power (CHP) and commercial electricity-only plants.
7 Electricity-only and combined-heat-and-power (CHP) plants whose primary business is to sell electricity, or electricity and heat, to the public. Includes 0.1 quadrillion Btu of electricity net imports not shown under "Source."

- The industrial sector is the single most important energy sink in the U.S., accounting for 31 percent of the total primary energy consumption. Close second is the transportation sector with 28 percent, followed by residential with 22 percent, and the commercial sector consuming the remaining 19 percent.
- Almost all the energy consumed in the transportation sector (93 percent) comes directly from petroleum.

Figure 8.1 shows how slanted the current economy is towards fossil fuels, especially for the transportation and residential/commercial sectors, whereas renewables accounts for less than 10 percent of the energy consumed.

Personal transportation represents all travel by individuals, including trips for shopping and leisure as well as commuting to work. It accounts for 8.9 million barrels of oil per day out of a total of 13.5 million barrels consumed, or approximately 66 percent (U.S. Energy Information Administration 2010). Together, personal transportation and residential energy consumption account for over one third of the total U.S. energy consumption. Thus, any improvement in energy conservation and efficiency in the residential and personal transportation sectors will directly and significantly affect the national energy situation.

The U.S. Energy Information Administration publishes yearly projections labeled "business-as-usual trend estimates", given known technology and technological and demographic trends. According to such projections,

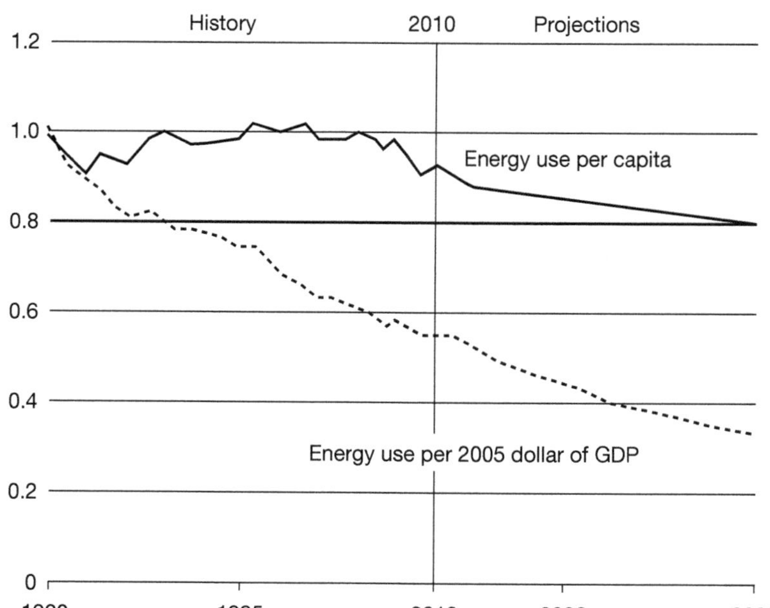

Figure 8.2 Energy Use per Capita and per Dollar of GDP, 1980–2035.
Source: EIA Annual Energy Outlook 2012.

overall U.S. energy consumption is expected to grow at an average annual rate of 0.3 percent from 2010 through 2035. The EIA does not expect the U.S. to return to the levels of energy demand growth experienced in the twenty years prior to the 2008/2009 recession, because of more moderate projected economic growth and population growth coupled with increasing levels of energy efficiency. For some end-uses, current federal and state energy requirements and incentives play a continuing role in requiring more efficient technologies. Energy consumption per capita is actually expected to decline by an average of 0.6 percent per year from 2010 to 2035 (see Figure 8.2). As shown in Figure 8.2, the amount of energy consumed per capita has been fairly steady since about 1970, with a slight decline in recent years. Even though we are living in larger homes, driving more miles, and using more electrical devices, we are actually consuming less energy because of improvements in efficiency and energy conservation techniques (this is especially true for passenger vehicles). Many of these improvements are directly attributable to public policies mandating increased energy efficiency. Additionally, the energy intensity of the U.S. economy, measured as energy consumption per real dollar of GDP, has steadily declined and is projected to decrease by an average of 2.1 percent per year from 2010 to 2035. This indicates better energy efficiency in all segments of American life, and a transition from an energy-intense manufacturing industry to less energy-consuming industrial and commercial processes. Obviously, however, there is still much room for improvement.

Energy Consumption in Urban and Rural Areas of the United States

Both residential and personal transportation energy consumption can be divided between urban and rural areas. Urban areas include all urbanized areas (over 50,000 population) and Urban Clusters (2,500 to 49,999 population) as defined by the Bureau of the Census in the 2000 Decennial Census. Ninety-two percent of all American households are considered urban by the Labor Department, compared to just 8 percent of households that are considered rural. Nevertheless, 19 percent of the U.S. population lives in rural areas (Census 2010 Population Statistics). Significant differences are noticed between these two settings, ranging from spending patterns, education, living habits, and energy consumption.

Residential and transportation energy consumption results from a combination of infrastructure and behavior factors, which make rural and urban areas quite different. Data from the U.S. Census Bureau show differences in household expenditures for rural and urban areas, especially regarding housing-related expenditures. The costs faced by families living in urban areas are generally higher, compared with their rural counterparts.

The U.S. Energy Information Administration periodically administers the Residential Energy Consumption Survey (RECS) to a nationally representative sample of housing units. Specially trained interviewers collect energy characteristics on the housing unit, usage patterns, and household demographics. This information is combined with data from energy suppliers to these homes to estimate energy costs and usage. All housing units in the fifty states and the District of Columbia that are occupied as primary residences are eligible to be included in the RECS sample, and a multi-stage design is applied to assure that the entire population of occupied housing units in the United States is represented. RECS is unique in the sense that it is the only survey that provides reliable, accurate and precise trend comparisons of energy consumption between households, housing types, and areas of the country.

Table 8.1 reports a summary of households' characteristics and energy use in urban and rural areas, as reported in the RECS data. Residential per household energy consumption in rural areas is almost 10 percent higher compared to those in urban areas, while electricity consumption is 50 percent higher.

Also, rural households tend to be larger (about 30 percent bigger compared to urban households) and newer (by about ten years). Even though more energy is used in rural areas, the proportion of energy devoted to various end-uses (five categories are proposed in the RECS: space heating, air conditioning, water heating, refrigerators, and other) is quite similar, as shown in Figure 8.3.

In the United States the majority of residential energy consumption is related to space heating, closely followed by other consumption, including refrigerators, lighting, and other appliances.

Table 8.1 Rural and Urban Households' Characteristics and Energy Consumption

	Urban	*Rural*
Number of households	92%	8%
Population	79%	21%
Average household construction year	1969	1980
Average household square footage	2045	2659
Average per household total energy consumption [kBTU/year]	88501	95944
Average per household energy consumption for heating [kBTU/year]	37059	36089
Average per household energy consumption for AC [kBTU/year]	5355	7322
Average per household energy consumption for water heating [kBTU/year]	16547	15145
Average per household energy consumption for refrigerators [kBTU/year]	4100	4806
Average per household energy consumption for other purposes [kBTU/year]	25440	32584
Average per household total energy related expenditure [$]	1952	2373
Average per household heating related expenditure [$]	537	648
Average per household air conditioning related expenditure [$]	193	241
Average per household water heating related expenditure [$]	268	329
Average per household refrigerators related expenditure [$]	153	157
Average per household expenditure related to other energy use [$]	800	999

Source: EIA RECS

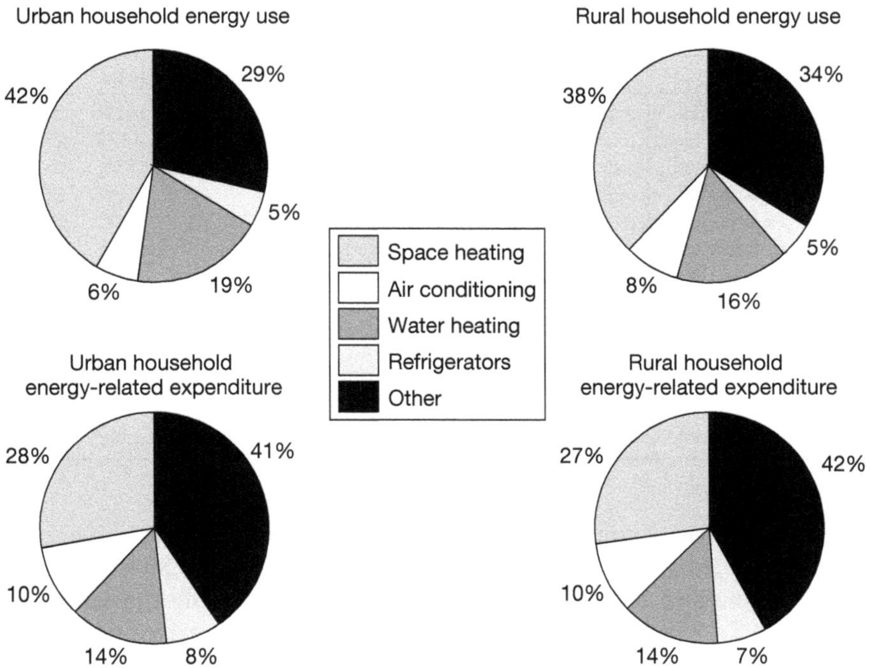

Figure 8.3 Urban and Rural Energy Use and Expenditures per Categories.
Source: EIA RECS.

The average annual energy-related expenditure tells another story, though. Rural families spend about 20 percent more that their urban counterparts, resulting in an average of an additional $400 per year per household. This difference is justified by the different sources of energy used. Even so, the proportion among the various categories does not change significantly in the two environments.

Figure 8.3 is also a useful indicator of different energy costs. The 'other' category is almost entirely represented by electricity consumption, and its weight on the total expenditure is significantly higher compared to the actual amount of energy consumed. Often natural gas, oil derivatives, or wood are used for space heating, which present a significantly lower cost.

Table 8.2 reports annual electricity consumption (divided by categories) as well as fossil fuels consumption for urban and rural dwellings.

Urban households tend to consume substantially more natural gas, while wood and heavy fossil fuels (i.e. oil and kerosene) are predominantly used in rural environments. Natural gas is largely used for space heating, where available. A lack of use of natural gas in rural areas can be directly related to a lack of infrastructures. The difference in price and available technologiesfor these different energy carriers leads to the aforementioned difference in energy-related expenditures, which is almost double the total energy consumption difference.

Table 8.2 Comparison of Urban and Rural Energy Consumption

	Urban	*Rural*
Average per household electricity consumption [kWh/year]	10290	15259
Average per capita electricity consumption [kWh/year]	4795	6766
Average per household electricity consumption for heating [kWh/year]	870	1511
Average per household electricity consumption for AC [kWh/year]	1569	2146
Average per household electricity consumption for water heating [kWh/year]	858	1868
Average per household electricity consumption for refrigerators [kWh/year]	1201	1408
Average per household electricity consumption for other purposes [kWh/year]	5790	8325
Average per household natural gas consumption [ft^3/year]	45722	19879
Average per household natural gas consumption for heating [ft^3/year]	27993	12642
Average per household propane consumption [gal/year]	8.9	165
Average per household fuel oil consumption [gal/year]	39.7	56.5
Average per household kerosene consumption [gal/year]	0.9	4.2
Average per household wood consumption [kBTU/year]	2246	13504

Source: EIA RECS.

The total electricity consumption is 50 percent higher for rural households compared to urban areas (15259 kWh/year compared to 10290 kWh/year), while the number of household members remains basically the same (an average of 2.69 people per household for rural areas, compared to 2.66 in urban areas). Therefore, the per capita electricity consumption shows a significant difference. This is due to several reasons, most evidently a higher square footage (rural households are on average 30 percent larger than urban households) and the lack of alternative energy carriers (e.g. natural gas). Furthermore, other external factors play a significant role.

As Stephanie Battles, director of the U.S. Energy Information Administration's Energy Consumption Division,[1] said:

> There are a lot of things that go into it. We know that urban areas are heat islands, for example. The temperature in the summertime is always higher [in urban areas], so they use more air conditioning. But in the wintertime, urban areas are also warmer, so they use less heat than rural areas.
>
> ...the compact construction of urban condo towers and apartment buildings helps insulate their indoor climates, while large homes common in less dense areas need more energy for heating and cooling, and have a harder time keeping air from leaking outside. It's also behavioral. For example, people in New York City are gone a lot, but people in rural areas, a lot of times they're home more often.

Thus, the differences in energy consumption are really the result of differences in both infrastructure and behavior.

The higher electricity consumption in rural houses, both per household and per capita, is ubiquitous among the different consumption categories (see Table 8.2). Even though comfort levels, captured as set temperature in the household, do not show significant differences between urban and rural areas, the higher square footage of rural households and the lack of easily available natural gas directly translate into higher consumption for heating and air conditioning.

Moreover, the data might indicate a tendency toward higher electricity consumption in rural areas, where average incomes are slightly higher compared to urban areas and housing expenditures are significantly lower, leaving room for higher electricity consumption. Again, however, the proportion of electricity among the five categories shows no significant difference between urban and rural houses, as shown in Figure 8.4.

Generally, more energy is consumed in rural areas to move people around compared to urban areas. This is mainly due to the longer distances driven and the lack of alternatives in rural areas, where public transportation is impractical. According to EIA data, each American urban household owns an average of 1.8 vehicles, compared to 2.2 for each rural household. Urban families also drive about 7,000 fewer miles annually than their rural counterparts, which results in an annual average saving of more than 330gallons of gasoline (assuming an average of 21 miles per gallon). Even if heavy urban traffic might lead to higher fuel consumption,[2] on average the energy cost for transportation is significantly higher for rural areas.

Urban residents tend to have greater numbers of options for personal transportation, including: public transportation, walking, and cycling. They also tend to need to travel shorter distances, allowing people to commute by bike or walk (also carpooling is possible, since people live close to each other).

Table 8.3 summarizes some transportation statistics collected by the EIA in their 2001 Transportation Energy Consumption Surveys (available online at www.eia.gov/emeu/rtecs/contents.html).

For urban drivers the average fuel economy is 20.53 miles per gallon, whereas for rural drivers the average fuel economy is 19.45 miles per gallon.

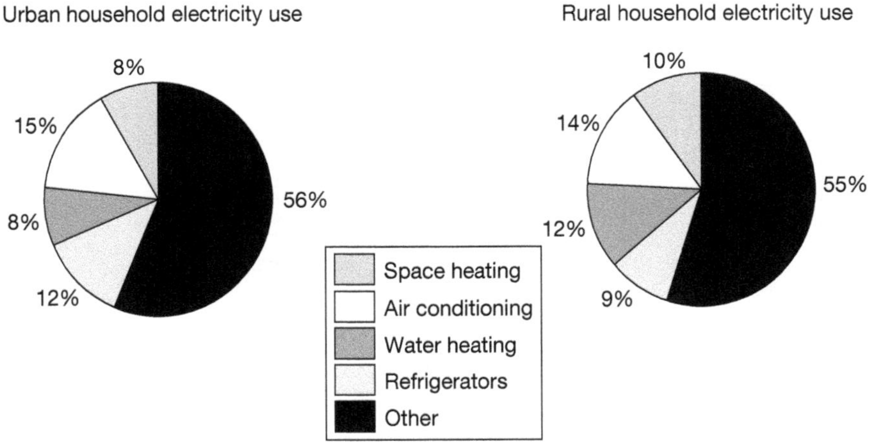

Figure 8.4 Electricity Use per Category in Urban and Rural Households.
Source: EIA RECS.

This indicates that although fuel economy for highway driving under idealized testing conditions is typically higher, in reality, rural drivers are less efficient with their transportation fuel than urban drivers. As opposed to households' ages, vehicles in rural areas tend be older, partially explaining the lower fuel economy. In addition, rural households have a higher number of vehicles per household, possibly to compensate for lack of other transportation options and longer distances.

In the two settings, different vehicle typologies are used for personal transportation. Figure 8.5 reports the breakdown per vehicle types in the two areas. As expected, more pick-up trucks and fewer passenger cars are used in rural areas. Such vehicles tend to be larger, more powerful, and therefore consume more fuel.

Figure 8.6 reports a comparison of the age of vehicles in rural and urban areas. Generally older vehicles are driven in rural environments, even though the difference does not seem to be dramatic.

Table 8.3 Comparison of Urban and Rural Personal Transportation Statistics

	Urban	*Rural*
Number of vehicles	75.1%	24.9%
Vehicle-miles traveled	73.3%	26.7%
Motor fuel consumption	72.3%	27.7%
Average per household number of vehicles	1.8	2.2
Average per household vehicle-miles traveled	21600	28700
Average per household fuel consumption	1054 gal	1469 gal
Average per vehicle miles traveled	11700	12800
Average per vehicle fuel consumption	570 gal	658 gal

Source: EIA RTECS.

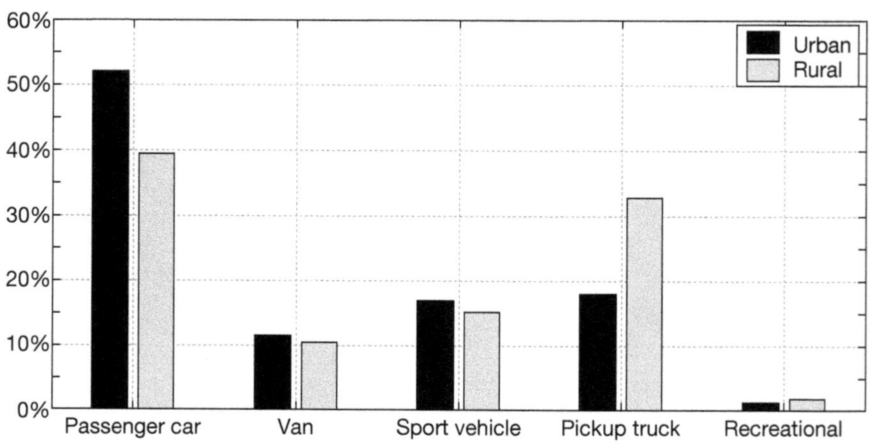

Figure 8.5 Statistics on Vehicle Type.
Source: EIA RTECS.

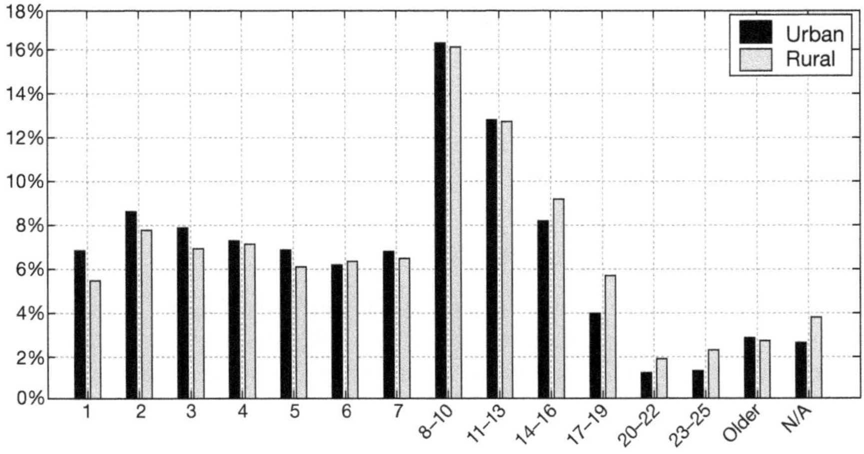

Figure 8.6 Statistics on Vehicle Age.
Source: EIA RTECS.

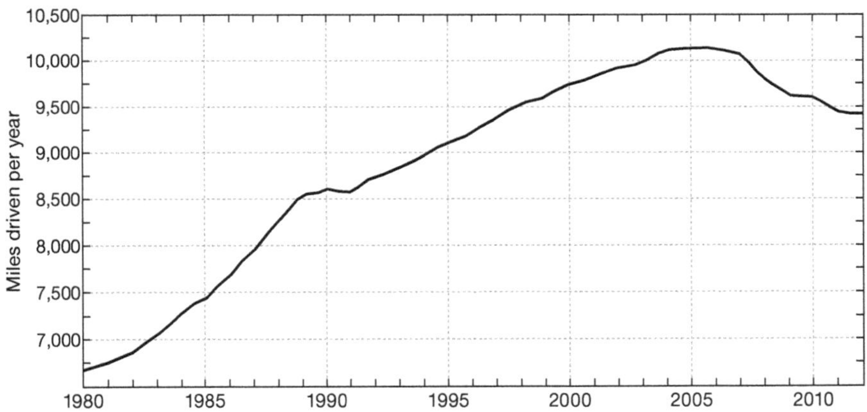

Figure 8.7 U.S. Miles Driven per Capita, Calculated by Dividing U.S. DOT Federal Highway Administration Estimates of Vehicle Miles Traveled (both cars and trucks) by U.S. Population, 1980–2012.

Looking at driving trends in the personal transportation sector, the number of miles driven annually by American citizens has been declining in the last five or six years, mostly due to the economic recession and increases in fuel prices (see Figure 8.7). This, combined with newer and stricter fuel economy standards and more environment-friendly regulations, should contribute significantly to reducing the total energy consumption of the personal transportation sector.

Automotive companies and researchers throughout the world have been putting a lot of effort into reducing the emissions and fuel consumption of conventional vehicles, as well as developing advanced vehicles—such as hybrid (electric/gasoline) vehicles, plug-in electric vehicles, hydrogen vehicles, natural gas vehicles, vehicles powered with bio-fuels or recharged with photovoltaic panels, autonomous vehicles, and advanced communications (vehicle-to-vehicle or cloud)—in an attempt to reduce the total energy and environmental footprint of personal mobility, without compromising the driver experience.

Energy Conservation and Efficiency Measures in Rural Areas

Rural households have substantial potential for increased energy efficiency and conservation. Vast amounts of energy could be saved if vehicles were driven fewer miles, and if homes used less energy. There are numerous policy and technology solutions that can help address these issues, including mandating better gas mileage for cars, providing incentives to increase the market penetration of advanced, more efficient vehicles, improving the insulation in homes, increasing the energy efficiency of appliances, the adoption of higher efficiency lighting, and others.

Smart Grid[3] technologies could also be used to better manage the use of electricity to improve the overall system efficiency. The Smart Grid will lead to demand response[4] programs, allowing private customers to modify their electricity demand profiles to fit the characteristics of flexible energy supply. This enables energy consumers to participate actively in energy markets (Schneider Electric, Demand Response and Smart Grid Services, n.d). In such a scenario, the entire energy grid becomes more efficient, costs less to operate, and presents a smaller environmental impact (Smart Energy Demand Coalition 2011). Still, to make this revolution possible, a system of over one million megawatts (solely in the United States [U.S. Energy Information Administration 2012]) will require a trillion-dollar retooling in the span of the next several decades (Fox-Penner 2010).

None of these solutions can address the disparity between rural and urban energy use alone, though, and it is most important that the intrinsic differences between the two settings must be considered while developing energy policies and energy-efficiency programs.

The greatest challenge is that these disparities are not simply the result of ignorance or lack of opportunity. People often live in rural settings by choice and they accept the trade-off in transportation time and energy-related costs in order to live the rural lifestyle. This includes the opportunities to live in larger homes. While policymakers can certainly develop policies to encourage greater conservation and mandate improved energy efficiency for new appliances and vehicles, often the adoption of these is dependent on economic incentives. Currently, the electricity market in the United States is structured in such a way so as to make it difficult to use electricity pricing to encourage conservation and the adoption of more efficient appliances.

Several incentives are currently available in the United States to help promote greater energy efficiency, including the adoption of high efficiency appliances (energy star rated appliances) and fluorescent/LED lighting. A 30 percent federal tax credit is available for the installation of geothermal heat pumps, wind turbines up to 100kW, solar water heaters, and solar PV systems through 2016. Moreover, states are promoting energy efficiency in several ways. Different electric and gas utilities offer a range of incentives for energy audits, energy efficiency, and renewable energy on top of federal and state tax credits and rebates (Weiner 2013).

However, greater efficiency is not always accompanied by lower energy use. For instance, the square footage of households plays a crucial role, and homes built since 1990 are on average 27 percent larger than homes built in earlier decades. In addition, while appliances are more efficient on average now than they were ten years ago, households often have a greater number of electrical devices.

More importantly, the cost of energy is still very low in the United States and many national policies are designed to keep it this way. Energy prices may actually discourage conservation and energy efficiency or the adoption of high-efficiency appliances, heating and air conditioning systems, and vehicles.

Energy efficiency and conservation are typically dependent on using existing energy resources better—that is, using less energy while still maintaining the same quality of living. For rural residents, the quality of living may be inherently more energy intensive. Furthermore, policymakers are presented with additional challenges in trying to promote conservation and efficiency: it is difficult to present economical alternatives in rural communities. Public transportation options are necessarily dependent on having enough users to support the transportation system. Rural areas have substantially lower population densities, making it difficult to justify public transportation, not to mention the fact that this lower density makes it difficult to design effective routes. Thus, rural residents are faced with fewer options.

In addition, many policies that provide assistance for improved energy efficiency and conservation are likewise intended to reach as many people as possible and thus are often tailored for urban residents. The energy efficiency programs that targeted poorer citizens for improved weatherization, improved insulation, and improved windows and doors were necessarily concentrated in urban settings, where there are a greater number of people that can benefit from these programs. It is difficult to justify these programs in low density regions. Moreover, few rural governments have the resources necessary to support energy conservation and efficiency programs since they have such limited resources in general. This disadvantages rural communities and residents with respect to energy conservation and efficiency, when they could benefit substantially from these programs.

Nonetheless, there are options for improving energy efficiency and conservation. Energy conservation does not necessarily require significant financial investments. For instance, driving styles substantially influence fuel economy. Aggressive driving styles, such as driving and accelerating fast, can reduce fuel economy by up to 10 percent (Vermeulen 2006). Also, education is an important avenue for promoting energy efficiency and conservation. Simply understanding the options available for reducing energy consumption and costs may help.

Conclusions

Energy is a key component of people's lives, and it has major socio-economic bearings. Promotion of energy conservation and efficiency measures in the residential and transportation sectors heavily impact the life habits of the residents. Also, energy improvement investments are expected to expand job opportunities. "This energy efficiency effort can help rural Americans reduce energy costs while simultaneously expanding business and investment opportunities in rural communities," said Agriculture Secretary Tom Vilsack (U.S. Department of Agriculture – Office of Communications (202) 720—4623, 2012).

When dealing with energy conservation, two main approaches are possible: increase person density (i.e. smaller households and public transportation/carpooling); or increase efficiency of energy use (i.e. high-efficiency appliances and lights, and improved vehicles).

Currently, the energy policies promoted in the United States are trying to increase person density in the personal transportation sector, while there are attempts at improving energy efficiency in both the transportation and residential sector. Nevertheless, there are significant differences when comparing urban and rural energy consumption, both for transportation and residential consumption.

First, different attitudes towards energy use and different needs are noted, especially for the transportation sector, where sparsely populated areas present particular characteristics and public transportation is impractical. Rural families drive more miles and generally use larger vehicles. On average rural households, which are newer, consume more energy. Nevertheless, they are generally more energy efficient per square foot than urban households, meaning that improved energy efficiencies may be more difficult to achieve than in urban households. However, energy comes in different forms, and rural households tend to consume substantially more electricity.

Second, the majority of energy is consumed in urban areas, even if per capita consumption in rural areas is higher. However, solutions involving increases in person density—which significantly affects lifestyle—do not apply to rural areas. Mass public transportation is unfeasible. Rural residents often choose to live in rural settings, where they can acquire larger properties, and they accept the trade-off of longer commutes to work. There are physical and geographic realities that make rural lifestyles more energy intensive. In addition, rural households have fewer energy options. Natural gas is less common for space heating, likely because the natural gas infrastructure is not as well developed in rural areas. The same happens for transportation, with a lot of focus given to highly efficient passenger vehicles, natural gas vehicles, and electric cars. Because of different final purposes, lack of infrastructure, and limited driving range, these vehicles do not cope well with rural needs for transportation.

Nonetheless, smart policies and innovative technologies can be used to reduce the overall energy consumption of a nation. Often a lot of attention is given to electricity and gasoline consumption, while a complete picture of residential and transportation energy consumption should be analyzed to optimize and rationalize the overall national energy use. In order to promote energy efficiency and conservation in rural and urban settings equally, further investigation of technological measures and policies is needed.

Acknowledgement

The author would like to thank Prof. Beth-Anne Schuelke-Leech for her constant support and for the valuable contributions provided during the preparation of this chapter.

Notes

1 Reported by Russell McLendon, available online at: www.mnn.com/earth-matters/translating-uncle-sam/stories/urban-or-rural-which-is-more-energy-efficient.

2 Fuel consumption is typically measured in miles per gallon, and is referred to as "fuel economy". Highway driving tends to yield better fuel economy because energy is not wasted in the frequent stops and starts required in city driving.

3 Smart Grid generally refers to a class of technology that people are using to bring utility electricity delivery systems into the twenty-first century, using computer-based remote control and automation. These systems are made possible by two-way communication technology and computer processing (U.S. Department of Energy – Office of Electricity Delivery & Energy Reliability, n.d.).

4 Demand Response is defined as: "changes in electric usage by end-use customers from their normal consumption patterns in response to changes in the price of electricity over time, or to incentive payments designed to induce lower electricity use at times of high wholesale market prices or when system reliability is jeopardized" [5].

References

Fox-Penner, P. 2010. *Smart Power – Climate Change, the Smart Grid, and the Future of Electric Utilities*. Washington, D.C.: Island Press.

Schneider Electric, Demand Response and Smart Grid Services. N.d. Online Available at: www.schneider-electric.us/sites/us/en/solutions/demand-response-solutions/demand-response-solutions.page (accessed 31 March 2014).

Smart Energy Demand Coalition. 2011. The Demand Response Snapshot – The Reality for Demand Response Providers Working in Europe Today. Online.

Available at: www.sedc-coalition.eu/wp-content/themes/blue-grace/images/snapshot.pdf (accessed 7 April 2014).

U.S. Department of Agriculture – Office of Communications (202) 720–4623. 2012. Online. Available at: www.usda.gov/wps/portal/usda/usdahome?contentidonly=true&contentid=2012/07/0241.xml (accessed 31 March 2014).

U.S. Department of Energy – Office of Electricity Delivery & Energy Reliability, n.d. Online. Available at: http://energy.gov/oe/technology-development/smart-grid (accessed 31 March 2014).

U.S. Energy Information Administration 2012. *Annual Energy Outlook 2012*, s.l.: s.n.

U.S. Energy Information Administration 2010. *Annual Energy Review 2009*, s.l.: s.n.

Vermeulen, R. 2006. *The effects of a range of measures to reduce the tail pipe emissions and/or the fuel consumption of modern passenger cars on petrol and diesel.*, s.l.: s.n.

Weiner, C. 2013. An Introduction to Rural Energy Options. Fort Collins, CO. Colorado State University Extension. Available at: www.ext.colostate.edu/pubs/consumer/10626.html (accessed 23 June 2014).

Index

f after a page number indicates a figure
t after a page number indicates a table

For Product Safety Concerns and Information please contact our EU representative GPSR@taylorandfrancis.com
Taylor & Francis Verlag GmbH, Kaufingerstraße 24, 80331 München, Germany

www.ingramcontent.com/pod-product-compliance
Ingram Content Group UK Ltd.
Pitfield, Milton Keynes, MK11 3LW, UK
UKHW011013180926
475701UK00004B/724

9781138240759